Genetics and Epigenetics of Breast Cancer

Madhumita Roy • Jaydip Biswas •
Amitava Datta

Genetics and Epigenetics of Breast Cancer

Springer

Madhumita Roy
Environmental Carcinogenesis & Toxicology
Chittaranjan National Cancer Institute
Kolkata, West Bengal, India

Jaydip Biswas
Surgical Oncology
Chittaranjan National Cancer Institute
Kolkata, West Bengal, India

Amitava Datta
Computer Sci and Software Engineering
University of Western Australia
Perth, WA, Australia

ISBN 978-981-19-9924-6 ISBN 978-981-19-9925-3 (eBook)
https://doi.org/10.1007/978-981-19-9925-3

This Springer imprint is published by the registered company Springer Nature Singapore Pte Ltd.
The registered company address is: 152 Beach Road, #21-01/04 Gateway East, Singapore 189721, Singapore

Preface

Breast cancer is one of the most common cancers that affects women. Millions of women worldwide are diagnosed with breast cancer every year, and the prognoses for many patients are not favourable in spite of improvements in treatments. Improved therapeutics depend on understanding breast cancer by classifying different types and subtypes, as treatments should be ideally very targeted for a patient. Hence understanding breast cancer is important both in terms of the comprehensive knowledge of the different types of breast cancer, molecular markers for the types and the genetic and epigenetic landscapes of the types of breast cancer.

This book is an attempt at concisely presenting the current understanding of breast cancer both from molecular biology and treatment perspectives, with an emphasis on genetics and epigenetics of breast cancer. The book is comprised of four chapters. The first chapter gives an overview of breast cancer based on epidemiology, types, molecular markers, signs and symptoms, diagnosis and treatments. The second chapter discusses in detail the genomic landscape of the different subtypes of breast cancer, starting from BRCA1 and BRCA2, the two genes that established the genetic susceptibility of breast cancer. This chapter also discusses the difficulties in determining the genomic landscapes precisely, due to heterogeneity of tumours and overlaps between different subtypes. Many of the drugs target the epigenetic modulators of breast cancer and the third chapter focuses the epigenetic landscapes of different subtypes of breast cancer. The epigenetic modulations include DNA methylation, histone modifications and post-translational silencing by miRNA. The fourth chapter discusses therapy and control of breast cancer in detail. Side effects are major problems associated with any treatment regime of breast cancer and natural products have the advantage that they do not have any side effects. The use of natural products for prevention and treatment of breast cancer is also discussed in depth in this chapter.

The book will be of interest to students of molecular biology and genetics for understanding the current knowledge of genetics and epigenetics of breast cancer. Chapters 2 and 3 can be used independently in a course on genetics and molecular biology. The book will be of interest to medical professionals for broadening their understanding of the molecular basis of breast cancer. The book also will be of interest for readers who want to get an overview of all aspects of breast cancer studying independently.

The authors would like to thank Dr. Bhavik Sawhney from Springer India, for guiding this project with patience and understanding.

Kolkata, India Madhumita Roy
Kolkata, India Jaydip Biswas
Perth, Australia Amitava Datta

Contents

About the Authors

Madhumita Roy is currently the Assistant Director and Head of the Department of Environmental Carcinogenesis and Toxicology, Chittaranjan National Cancer Institute, Kolkata. Her research focuses on cancer biology, environmental toxicology, and identification of important tumor biomarkers, and chemo-enhancing efficacy of tumor cells by plant-derived molecules. She has published over 80 research articles in journals of high repute. She is a life member of several societies, such as the Environment Mutagen Society of India (EMSI), Indian Association for Cancer Research (IACR), Electron Microscope Society of India, Indian Science Congress Association, All India Congress of Cytology and Genetics, and the Society of Nutraceuticals and Chronic Diseases.

Jaydip Biswas completed his MBBS, MS (Master of Surgery in general surgery), and Diploma in Medical Therapeutics from the University of Calcutta, FRCS Ed from the Royal College of Surgeons of Edinburg, FACS from the American College of Surgeons, and PhD in Medicine from the University of Calcutta. He has worked in various roles in several hospitals, and in particular at the Chittaranjan National Cancer Institute, Kolkata, where he was an officer in charge, surgeon, and director between 2013 and 2018. He has received further medical training at Memorial Sloan Kettering Cancer Centre, New York, and Manchester University Hope Hospital, Salford, UK, and Christie Hospital, UK, in surgical oncology. He has extensive clinical expertise in different types of elective, emergency general surgical procedures including different types of cancer surgery, head and neck reconstructive procedures, breast reconstruction, and GI endoscopy. He has taught many courses on surgical oncology at Chittaranjan National Cancer Institute. He has extensive research experience in various cancers and has published over 70 papers in leading national and international journals on cancer, specifically focusing on the role of phytochemicals in cancer treatment. He has also conducted many phase II and phase III clinical trials for potential therapies by drugs and phytochemicals in various cancers.

Amitava Datta is currently a professor in the Department of Computer Science and Software Engineering at the University of Western Australia. He has completed his MTech and PhD in Computer Science at the Indian Institute of Technology, Madras.

He pursued postdoctoral research at the Max Planck Institute for Computer Science and the University of Freiburg, both in Germany. He was a visiting professor at the University of Freiburg and is currently a visiting professor at the University of Life Sciences, Warsaw, Poland, and South West University, China. His scientific interests include computational molecular biology, data mining, and parallel and distributed computing, and his computational molecular biology research focuses on RNA structural biology and miRNA gene and target prediction, particularly in relation to cancer biology. He has published over 150 research papers in respected international journals.

Breast Cancer: Epidemiology, Types, Diagnosis, and Treatment

1

1.1 Introduction

Breast cancer is one of the most common cancers that affects women. Millions of women worldwide contract this disease every year, and the survival rate depends on proper diagnosis, treatment, and support. Moreover, the recurrence of cancer after initial treatment and eradication of a primary tumour is not uncommon. Hence it is of utmost importance to improve our knowledge of breast cancer from all aspects to improve the treatment of this disease. Cancer is a molecular disease; the initiation of cancer starts from genetic and epigenetic abnormalities in cells. Our understanding of the genetic landscape of breast cancer is improving significantly due to intensive research efforts, however genetic abnormalities or mutations are usually irreversible and cannot be changed by treatments. The only option is to destroy the cells by applying radiation. Epigenetic changes on the other hand can be reversed through drugs, and a major focus of research in breast cancer is to discover proteins that are responsible for epigenetic abnormalities and can be targeted through drugs. Moreover, understanding the molecular subtypes of breast cancer is very important for deciding on an appropriate treatment plan. Another important aspect of research is to identify biomarkers of breast cancer, molecules whose presence can give accurate clues for the type and stage of a cancer, so that it can be treated appropriately. We discuss all these aspects in this chapter, along with other important aspects like signs and symptoms of breast cancer and the available treatment options.

1.2 Epidemiology of Breast Cancer

Epidemiology is the area of study of the incidence, causes, and distribution of health-related events or disease states in a population. An epidemiologist therefore is involved in the occurrence, distribution, and control of different diseases. Ultimate aim of this branch of science is to provide necessary guidelines for controlling health

M. Roy et al., *Genetics and Epigenetics of Breast Cancer*,
https://doi.org/10.1007/978-981-19-9925-3_1

problems. This is true for the dreadful disease cancer as well. Prevention and awareness are the key to cancer control, as they can reduce the incidence, mortality due to cancer and hence have important effects on the reduction of the burden due to cancer. Cancer is one of the leading causes of death and a hurdle that restricts life expectancy. The number of deaths due to cancer has an upward trend in spite of worldwide research efforts. Moreover, the quality of life, even after recovery, is often compromised for a cancer patient.

There are 10 hallmarks of cancer (Hanahan and Weinberg 2011; Roy and Datta 2019), among which evasion of apoptosis, uncontrolled proliferation, augmentative angiogenesis, defiance to anti-growth signals and evocation of own growth signals, and the ability to spread to distal parts, i.e. to metastasize (Łukasiewicz et al. 2021) are critical.

Global Cancer Observatory (GLOBOCAN) gives an estimate of cancer incidence, mortality, and prevalence around the globe. International Agency for the Research on Cancer (IARC) released the latest GLOBOCAN 2020 report and as per the latest estimate, the incidence of cancer has risen to 19.3 million cases, resulting in 10 million deaths (UICC 2020). According to the latest GLOBOCAN 2020 report, among all cancers, prevalence of breast cancer is the highest among women; there are 2.3 million new cases around the globe every year (Sung et al. 2021). Incidence of colorectal, lung, cervical, and thyroid cancers is also not to be ignored among women. In male population, incidence of lung cancer is the highest. Incidence wise prostate and colorectal cancers come next, whereas mortality due to liver and colorectal cancers is quite high. According to IARC, 1 in every 5 people irrespective of their gender develops cancer during their lifetime. Of these, incidence of death is 1 in 8 in the male population and 1 in 11 in the female population. This book is focussed on breast cancer as it is the most common cancer among women.

According to GLOBOCAN 2018, ASIR (age-standardised incidence rates) of this cancer directly correlated with the Human Development Index (HDI) (Sharma 2021). HDI is a statistical index to measure socio and economical dimension of a country, in terms of life expectancy, education, and per capita income indicators; this index ranks countries into different tiers of human development. According to GLOBOCAN 2020, the ASIR is directly proportional to the HDI, e.g. it is highest in countries having high HDI index, whereas for countries with medium and lower HDI, ASIR is relatively lower (Ferlay et al. 2020).

Breast cancer forms in the cells of the breasts; this cancer starts in the epithelial lining of the ducts or lobules in the glandular tissue of the breast. At the inception, malignant growth is restricted to the duct or lobule, which is called "in situ" carcinoma. The term "in situ" means "in its original place" and is used to define a cancer which is confined within the cells where it started. With passage of time, this in situ or localised growth may progress and invade the surrounding tissues, which is called invasive breast carcinoma. With further progression, the offending growth may invade the nearby lymph nodes and finally spread to distal organs of the body; these spreads are coined regional metastasis and distant metastasis, respectively. It can occur in both men and women, though breast cancer is common among women, rather than among men. Occurrence of breast cancer in men is less than 1% of all occurrences, but outcome may not be good, due to its late diagnosis among males. Irrespective of the genders, human breasts comprise nipples, fatty tissue, breast cells, and ducts. There are some common risk factors among both the genders. Risk of cancer may be due to some inherited mutations in the BRCA1 and BRCA2 genes, in both men and women. Production of the hormone oestrogen in both sexes (more in female, less in male) may contribute to the risk of breast cancer. However, there are a few differences, male breast is comprised of mainly fat and fibrous tissue called stroma with fewer ducts and lobules. On the other hand, for women, with age, particularly during puberty, working lobules and milk ducts develop, which help in lactation. Most of the female breast cancers develop in ducts and lobules.

Breast cancer is not a transmissible or contagious disease. Therefore, in terms of its aetiology, breast cancer is heterogeneous with a number of pathological characteristics. There is no single cause, and therefore aetiology of breast cancer is multifactorial. Some of the risk factors are avoidable, some like genetic predisposition are beyond control. However, it is important to know the aetiology of breast cancer, so that we can learn from these causes, reducing the risk. Controllable risk factors include alcohol consumption, body weight, lack of physical activity, pregnancy, no breastfeeding, smoking, consumption of contraceptive pills, and hormone replacement therapy. The factors that cannot be controlled comprise gender, breast density, age, age at menarche and menopause, reproductive history, exposure to radiation, family history of breast cancer, previous history of breast cancer, and use of diethylstilbestrol (DES).

For a better understanding of the causes of breast cancer, the points mentioned above need to be elaborated. Drinking alcohol increases the risk of breast cancer. Individuals who consume alcohol even in moderation have a higher chance of contracting breast cancer compared to a teetotaller. Obesity or over weight is another cause of development of breast cancer. This is again linked to oestrogen production; obesity after menopause produces more oestrogen, which may contribute to breast carcinogenesis. Epidemiologic evidence indicates a link between lack of physical activity and the likelihood of breast cancer. This may be due to the fact that increased physical exercise may reduce the risk of cancer by controlling adiposity, insulin resistance, sex hormones, adipokines, and chronic inflammation (Lynch et al. 2011).

Early detection and diagnosis are the best ways to prevent the disease. Therefore, it is of enormous importance to know the pathogenesis, genes involved, risk factors, and preventative methods of breast cancer (Sun et al. 2017). Reproductive history of a woman is an important determinant in breast cancer. Pregnancy reduces the risk of breast cancer. Abortions seem to have no impact on the risk. A study indicates that at least 34 weeks of pregnancy may be associated with reduced risk of breast cancer. Full term pregnancy at an early age and multiparity lower the risk of breast cancer (Husby et al. 2018). Nulliparous women also have a greater risk of breast cancer. Breastfeeding also has an impact on breast cancer risk. Risk is much lower if breastfeeding is continued for more than a year. Hormonal changes occur during and after pregnancy, which halt the monthly period. Breast feeding also impacts the hormonal changes. Therefore, during pregnancy and lactation, less exposure to hormone occurs. Oestrogen and some other hormones are associated with breast cancer risk.

Smoking is injurious to health; it is a known cause of cancer. Cigarette smoking releases a plethora of chemicals which are harmful for our health. Many of these chemicals are carcinogens, which enter the lung and then spread around the entire body. These harmful chemicals damage the DNA, and often cells fail to repair the damage, causing mutation and accumulation of many such mutations ultimately lead to cancer. Like many other cancers, smoking increases the risk of breast cancer. The risk is even higher with a family history of breast cancer.

Contraceptive pills slightly increase the chance of breast cancer. Within a few years of stopping, however, this risk disappears. A cohort study pointed to the risk of invasive breast cancer in women using hormonal contraception (Mørch et al. 2017).

Ovarian function is greatly compromised with the advent of menopause, as oestrogen and progesterone levels fall appreciably. Hormones being chemical messengers can lead to menopausal symptoms, and mood swings and hot flushes are quite common. Hormone replacement therapy (HRT) is used to tackle the symptoms of menopause. HRT types that are used are (a) only oestrogen and (b) a combination of oestrogen and progestogen. Long-term use of HRT increases the risk of breast cancer. However, the risk of breast cancer diminishes after cessation of HRT (Narod 2011; Beral and Million Women Study Collaborators 2003).

There are certain risk factors of breast cancer, which cannot be modified. First comes gender, being a woman is a potential risk factor. Men may have breast cancer, but the incidence is less than 1%, which is quite low. The reason could be that men's breast cells are inactive and have extremely low levels of oestrogen. For women, breast development is complete usually by the age of 14. Breast cells in women are immature till first full-term pregnancy. During this immature stage, breast cells are very responsive to oestrogen. Hormonal stimulation in women, particularly during breast developmental stage may be a reason why women are more prone to breast cancer than men. A pictorial representation of the risk factors, modifiable, as well as non-modifiable is shown in Fig. 1.1.

Breasts are composed of ducts, lobules and fatty, fibrous and glandular tissues. Lobules produce milk, and milk is transported from lobules to the nipple through the ducts. Fat and fibrous tissues are responsible for the size and shape of the breast. Dense breasts having more glandular and fibrous tissues than fat are more susceptible to breast cancer (Boyd et al. 2010).

Age is another important determinant of the risk of breast cancer. Glands in breast change with aging and menopause, and these changes put an individual at risk of developing breast cancer. Rate, extent, and type of damage to the DNA as well as the cell's ability to repair the damage determine the final consequences (Benz 2008). Most of the diagnosed women with breast cancer are older than 50. Therefore, getting older imparts a significant risk for developing breast cancer.

Menarche (first occurrence of menstruation) and menopause (end of menstrual cycle) are the beginning and end of ovarian activity. Reproductive period is the time between menarche and menopause, and during this time ovary starts producing steroid hormones that are involved in the development and function of the breast.

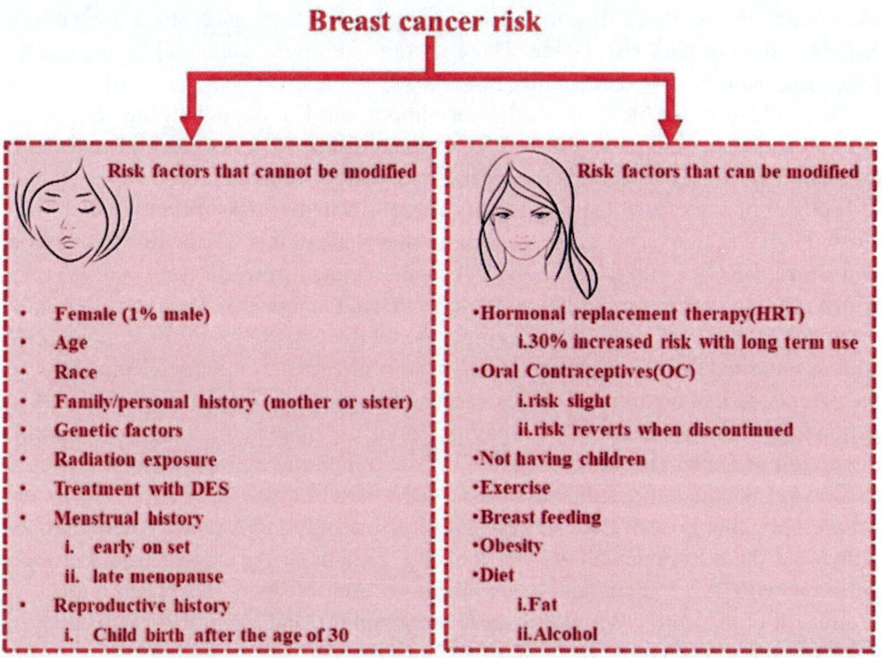

Fig. 1.1 The non-modifiable and modifiable risk factors of breast cancer are shown. Women are at elevated risk owing to factors that cannot be changed, i.e. non-modifiable risk factors. However, onset of breast carcinogenesis may be prevented by lifestyle changes, keeping in mind the modifiable risk factors

Early menarche and late menopause result in an increased span of reproductive period and more production of hormones. This increases the risk of breast cancer (Collaborative Group on Hormonal Factors in Breast Cancer 2012). Women who have menarche before the age of 12 and menopause after 55, and who have never been pregnant, have very high exposure to oestrogen and hence breast cancer risk is also very high for such women.

Race and ethnicity have roles in the risk of breast cancer. Risk of breast cancer in white women is higher than Black, Hispanic, and Asian women. Triple-negative breast cancer is an aggressive form of breast cancer, and it is more common among Black women. Death among Black women due to breast cancer is higher. The reason could be inaccessibility to mammography and poor medical care; lifestyle may have a role too (Chlebowski et al. 2005).

Radiation therapy (or radiotherapy) is a therapeutic modality that uses ionising radiation, delivered by a linear accelerator. High dose radiation is used to kill malignant cells and to shrink tumours. Radiation therapy to cure a different cancer increases the risk of breast cancer at a later time. Extent of risk is dependent on the age at which radiation exposure occurs: if exposure occurs before the age of 20, the risk is greater. There are other diagnostic procedures in which X-rays and CT scans are employed; they too add to the risk of breast cancer (Ronckers et al. 2005).

Mammography, used for diagnosis and screening of breast cancer, uses low energy X-rays and is another risk factor. There is a possibility of collateral breast cancer following radiotherapy (Eidemüller et al. 2021).

Family history for breast cancer is very important. Having a first-degree relative with breast cancer increases the risk factor many folds. Risk is even higher if more than one first-degree relative had cancer. Also, history of male breast cancer (father or brother) may increase the risk factor. There is a higher risk of breast cancer if a close relative is an ovarian cancer victim. In this context, it is worthwhile to mention two genes, BRCA1 and BRCA2. BRCA is abbreviated from BReast CAncer gene, which are responsible for producing proteins that aid in repair of damaged DNA, and these are known as tumour suppressor genes. Of the two copies of these genes, one copy is inherited from each parent. When a harmful variant is inherited from either of the parents, risk of breast and ovarian cancers may increase. Loss of BRCA1 leads to genomic instability, loss of DNA repair potential, anomaly at the transcription level, disruption of G2/M checkpoint regulation, etc., rendering an individual susceptible to cancer. Like the tumor suppressor gene TP53, BRCA1 also plays a role of a tumor suppressor gene. Loss of BRCA2 leads to impairment of DNA repair potential of an individual and hypersensitivity to ionising radiation. An association between germline BRCA2 variant and triple-negative breast cancer has been indicated (Petrucelli et al. 1998). We will discuss the genomic landscape of breast cancer in details in the next chapter, including BRCA1 and BRCA2.

A breast cancer survivor has a high risk of developing breast cancer again, compared to a person having no previous history. Someone having non-invasive cancer at early stage will be at risk of developing cancer in the same or in the other breast. A woman having ductal carcinoma in situ (DCIS) or invasive breast cancer is susceptible to develop a new breast cancer and the risk factor is higher. Even atypical hyperplasia, which is a non-cancerous breast disease, or lobular carcinoma in situ is associated with a higher risk of breast cancer.

Cancer in the first breast could be a risk factor for cancer in the other breast, which is called contralateral breast cancer (CBC). In a woman, having a previous or a cancer in the contralateral breast with a history of previous carcinoma of breast, cancer of contralateral breast cancer is regarded as an independent primary tumour (Begg et al. 2018). In a breast cancer diagnosed woman with BRCA1 or BRCA2 mutation, the risk of cancer in the contralateral breast is dependent on the first diagnosis. There are modifiers of cancer risk, both genetic and non-genetic (Narod et al. 2014).

Diethylstilbesterol (DES) is a synthetic hormone oestrogen. It is a chemical that disrupts endocrine function, may cause cancer, birth defects, and other abnormalities. It was used to prevent miscarriage, premature labour, and other complications of pregnancy in the past (Anonymous 1976). It was also used to halt lactation, as emergency contraception and also to treat menopausal symptoms (Al Jishi and Sergi 2017). Daughters of the mother, who took DES during pregnancy, developed problems like miscarriage, premature birth, and ectopic pregnancy. Prenatal hormone levels also increase the risk of breast cancer (Palmer et al. 2006).

1.3 Biomarkers in Breast Cancer

Biomarkers play an important role in the detection and treatment of breast cancer patients. Some of these have been discussed below, and some of these along with their functions have been summarised in Table 1.1.

There are certain endogenous proteins or metabolites called tumour cell markers, whose levels are indicative of the state, progression of tumour and its response to therapies. For surveillance of breast cancer some such markers CA 15-3, carcinoembryonic antigen (CEA), and CA 27-29 are used. Among the markers that are commonly used in breast cancer diagnosis are urokinase-dependent plasminogen activator system (uPA), plasminogen activator inhibitor (PAI) and Thomsen–Friedenreich (TF) antigen. Other biomarkers include mammaglobin, osteopontin, snail, twist, Zeb-1, fibroblast growth factor receptors (FGFR), phosphatase and tensin homolog (PTEN), and sirtuins (SIRT). PAM50, ROR, Breast Cancer Index, and EndoPredict are predictive of development of late recurrence, which is important for targeting patients for extended hormone therapy (Filipits et al. 2014).

Tamoxifen is a drug of choice in the treatment of breast cancer. In order to predict tamoxifen response, retinoic acid receptor alpha (RAR-α) may be a good option (Chung et al. 2014). STAT3 (Signal transducer and activator of transcription 3) is responsible for tamoxifen resistance in breast cancer (Chung et al. 2014). The tyrosine-kinase receptor has extensive homology with the epidermal growth factor receptor and is encoded by the oncogene HER-2/neu. For endocrine-sensitive breast cancer patients, anti-HER2 therapy is recommended. Evidence suggests that Herceptin (anti HER2 antibodies) in conjunction with chemotherapeutic drugs delays recurrences in metastatic breast cancer. Aberrant expression of STAT3 leads to tamoxifen-mediated apoptosis. Metastasis is a major problem in breast cancer, which is responsible for poor prognosis. A group of proteins termed matrix metalloproteinases (MMPs) regulate the metastatic events. A member of the tissue inhibitors of metalloproteinases (TIMPs) family, tissue inhibitor of metalloproteinase 1 (TIMP-1), controls the activity of MMPs and in addition aids in the regulation of cell proliferation, apoptosis, and angiogenesis. These functions imply that TIMP-1 may play a role in tumour development and growth. Elevated TIMP-1 therefore correlates with poor disease-free survival and overall survival in patients. TIMP-1 has gained enormous importance in cancer research due to its significant role in development of resistance to chemotherapy (Gong et al. 2014).

Mutation in the tumor suppressor genes BRCA (both BRCA1 and BRCA2) might often lead to hereditary breast cancer (Welcsh and King 2001). Four genes have been identified, namely CHEK2, ATM, BRIP1, and PALB2, which are functionally related to BRCA (Turnbull and Rahman 2008). The guardian of genome p53 is a key marker in carcinogenesis as it regulates cell proliferation and takes care of chromosomal stability. Mitogen-activated protein kinase (MAP kinase) signalling pathways also play pivotal roles in breast cancer. Hormonal regulation finds importance in pathology of breast cancer. Oestradiol, progesterone, and testosterone are known to activate MAP kinase pathways (Henrich et al. 2003). Some of the prognostic biomarkers of breast cancer are high level of proliferation indices such

Table 1.1 Biomarkers in breast cancer and their important functions

Breast cancer biomarkers	Functions
CA 15–3	A monitoring tumour marker considered for the prognosis of bone metastases in patients with breast cancer. This marker is highly relevant for treatment outcomes and disease recurrence
CEA	A family of cell surface glycoproteins, it is a monitoring marker in clinical practice in metastatic breast cancer
CA27.29	A monitoring marker, mainly in metastatic breast cancer. This cancer antigen can also predict early recurrence of cancer in women with treated carcinoma of breast
uPA	Diagnostic and prognostic marker in invasive BC. Urokinase plasminogen activator (uPA) is a serine protease that plays a role in both invasion and metastasis. This marker is a predictor in premenopausal women
PAI	PAI is an inhibitor of uPA, a prognostic marker in invasive BC
TF	Diagnostic and prognostic marker, an antigen which is aberrantly glycosylated carbohydrate, often found in adenocarcinomas. This antigen is a predictor of breast atypia and cancer in postmenopausal women
Osteopontin	A phosphorylated glycoprotein, which binds to cell surface integrins. This is implicated in invasion and metastasis, owing to its adhesive properties. Prognostic significance of this molecule is highly relevant in early breast cancer and progression. Osteopontin level correlates with relapse, cancer aggressiveness
Mammaglobin	This is expressed in normal, as well as in breast cancer cells. This biomarker is a diagnostic marker
Snail	Mesenchymal marker, it is a predictive biomarker for metastasis and invasion
Twist	Mesenchymal marker, it is a predictive biomarker for cell migration, metastasis, and invasion
Zeb	Mesenchymal marker, it is a predictive biomarker for metastasis and chemotherapeutic resistance
FGFR	Fibroblast growth factor receptor (FGFR) has a tyrosine-kinase intracellular domain. Role of FGFR2 polymorphisms is a marker for susceptibility to cancer, progression, and metastasis
PTEN	A tumor suppressor gene, which is deleted or mutated in many types of cancer. Loss of PTEN causes loss of apoptosis and aberrant cell cycle regulation. Also, loss of PTEN is correlated with increased tumour size. This is also a predictor of tamoxifen response
SIRT	Sirtuins (SIRTs) play an important role in tumour biology, particularly for breast cancer. Breast cancers have one deleted copy of this gene and are implicated in breast tumorigenesis and its progression and metastasis
Cytochrome P450	This is very important in prediction of systemic chemotherapy resistance. Breast cancer patients treated with tamoxifen often encounter relapse. CYPs serve a multi-faceted role in contributing to carcinogenesis, tumour growth, invasion, and metastasis
PIK3CA	This is also a predictor of resistance to chemotherapy, particularly to trastuzumab. PIK3CA is an alpha catalytic subunit alpha catalytic subunit in the PI3K/PTEN/Akt signalling pathway that controls cell Growth, proliferation, motility, survival, differentiation. Patients with HR positive, HER2 negative breast cancer, mutation in PIK3CA are common

(continued)

Table 1.1 (continued)

Breast cancer biomarkers	Functions
RARA	This biomarker is also an important predictor of tamoxifen response. Retinoic acid receptor alpha (RAR-alpha) is a known oestrogen target gene in breast carcinoma. Tamoxifen-resistant cells showed elevated levels of RARA
STAT3	Signal transducer and activator of transcription 3 (STAT3) is a transcriptional activator in *breast cancer*, which regulates several target oncogenes and affects *breast cancer* progression. This molecule is a proliferation inhibitor
TIMP-1	Tissue inhibitor of metalloproteinase 1 (TIMP-1) is a member of the inhibitors of metalloproteinases (TIMPs) family, which regulates the activity of matrix metalloproteinases (MMPs). TIMP-1 plays a vital role in tumour development and growth. This is also a predictor of response to chemotherapeutic drugs
Micro RNAs	These small non-coding RNA molecules play vital roles in regulation of gene translation. Among the many miRNAs identified so far, some are implicated in breast carcinogenesis. These are good predictors of disease progression and drug response to chemotherapy. Important among these are miR-21, miR-10b, miR155, miR-195, miR-222. miR-21 upregulation leads to resistance to trastuzumab via reduction of its gene target PTEN

as Ki-67 and proliferating cell nuclear antigen (Inwald et al. 2013). Breast cancer may be classified based on the expression of oestrogen receptor (ER), progesterone receptor (PR), HER2/ERBB2, transforming growth factor-α (TGF-α), and EGFR (Mohibi et al. 2011). All these markers together with bcl-2 overexpression are indicative of apoptotic imbalance. The decision of programmed cell death is finally governed by the bax/bcl-2 ratio. Oestrogen receptor (ER) through induction of anti-apoptotic proteins BCL-2 hinders apoptosis (Choi et al. 2001). Receptors play a dual role; some of the receptors like Erα, GRα, and PR promote proliferation, while ERβ and AR prevent cell proliferation (Khan et al. 2013). In case of various cancer conditions, within the tumour micro-environment, there has been a link between the inflammation and invasive potential, leading to poor prognosis of the disease. Cytokines like Interleukin-6 (IL-6), tumour necrosis factor alpha (TNF-α) and interleukin-1 beta (IL-1β) are mediators of inflammation. A link between inflammatory response, cytokines, and progression of breast cancer is well documented (Sirotkovic-Skerlev et al. 2012). TNF-α also activates nuclear factor kB (NF-κB) via canonical phosphorylation of NF-κB inhibitor kinase (IKK).

NF-κB, a transcription factor, is known to be associated with all the stages of breast cancer as well as response to chemotherapy and radiotherapy. NF-κB is an inducible transcription factor, regulating the expression of genes that control apoptosis and cell proliferation. This molecule is a heterodimeric complex of Rel family proteins (p50, p52, p65, cRel, and RelB). They confine themselves to the cytoplasm via interaction with inhibitor κB (IκB) proteins. In breast cancer, NF-κB consists predominantly of p65/p50 heterodimers (Cao and Karin 2003). Upon DNA damage or inflammatory cytokines, nuclear translocation of NF-κB occurs. In the nucleus, it binds to enhancer/promoter regions of target genes, activating transcription (Ghosh and Karin 2002).

Poly (ADP-ribose) polymerase (PARP) has a vital role in modulating the response to DNA damage (Veuger et al. 2009). The protein Poly (ADP-ribose) polymerase (PARP), a DNA repair enzyme is associated with various cellular process like genomic instability and apoptosis. This enzyme is responsible for detection and initiation of a number of responses involved in single strand repair (ssb). PARP has two isoforms, PARP-1 and PARP-2. Zinc finger DNA binding domains like XRCC1, DNA ligase III, DNA polymerase β are recruited for the repair. PARP-1 and PARP-2 oligomerise participate in base excision repair (BER). There is an overlapping role for PARP-1 and PARP-2 for these two isoforms to the damaged site of DNA. A number of other repair enzymes are also involved (Hanzlikova et al. 2017). PARP also plays a role in nucleotide excision repair (NER). PARP is involved in repair due to alkylating agents and chemotherapeutic drugs. PARP plays a role in restarting interrupted replication fork and BRCA1/2-mediated homologous recombination. Mutations in the BRCA1/2 genes result in hereditary breast cancer. These two tumor suppressor proteins participate in the homologous recombination repair of double stranded breaks (dsb) in DNA (Turk and Wisinski 2018). PARP-1 has a role as a transcriptional regulator with the enzyme that controls NF-κB. IR induced NF-κB activation has been found to be dependent on PARP-1 catalytic activity (Veuger et al. 2009). The fact that NF-κB activation gets abrogated by knockdown of PARP-1 supports these findings (Veuger et al. 2009). Irrespective of the signalling pathways through which NF-κB is activated, PARP-1 plays a vital role.

1.4 Types of Breast Cancer

Carcinoma of the breast can begin in different areas of the breast, e.g. the ducts, lobules, or in the tissues between the ducts and lobules. Type of breast cancer is determined by the kind of cells that become cancerous. Either one, or both the breasts may have cancer. The major classification is based on histopathological findings. Grade of tumour, its stage, along with the expression of genes and proteins aid in classification of breast cancer. Breast cancers are normally carcinomas. In a broader sense, breast cancers may be non-invasive, invasive, or metastatic, depending on their intrinsic and molecular subtypes. Different types of breast cancer are:

1. **Ductal Carcinoma in Situ (DCIS)**: DCIS, a non-invasive breast cancer starts inside the milk ducts and does not spread beyond the milk ducts to any adjoining tissues. It is localised. Though this is non-invasive in nature, may progress to invasive in nature later on.
2. **Invasive Ductal Carcinoma (IDC):** IDC, also known as infiltrating ductal carcinoma, is the most common type of breast cancer, which originates in the ducts. It can spread to other parts of the breast tissue, hence the name. This cancer

can penetrate the wall of the milk duct and invade the tissues of the breast and eventually can spread to the lymph nodes and other parts.

IDC Types

(a) **Tubular Carcinoma of the Breast:** It is a type of IDC, very small in size and is made up of tube-shaped structures, hence the name tubular. These cells resemble the features and characteristics of normal cells. This cancer is less aggressive, and its responsiveness to the treatment regime is good.

(b) **Medullary Carcinoma of the Breast:** This is another subtype of IDC, where the tumour is soft and fleshy and due to its resemblance to medulla, a part of brain, it is so named. Though this cancer may occur at any age, it is more common between late 40s and early 50s and in women with mutated BRCA1 gene. These slow growing carcinomas normally remain confined within the breast. Prognosis is not bad.

(c) **Mucinous, Papillary, and Cribriform Carcinoma of the Breast:** Mucinous, Papillary, and Cribriform Carcinoma are different types of IDC. In mucinous, also called colloid carcinoma, cancer cells float in mucin. Mucins are high molecular weight, glycosylated proteins produced by epithelial tissues. This type of cancer is not much aggressive and responds well to treatment.

Invasive papillary carcinomas are very rare and are seen in postmenopausal older women. An invasive papillary carcinoma has a well-defined border usually and comprises small, finger-like projections.

The cancer cells in invasive cribriform carcinoma invade the connective tissues of the breast called stroma. In the tumour, there are holes between the cancer cells, giving the appearance of cheese. In this type, cancer cells resemble the normal cells and this type of carcinoma is less aggressive.

3. **Invasive Lobular Carcinoma (ILC):** Invasive lobular carcinoma (ILC) is a common type of breast cancer, next to IDC. This cancer crosses the walls of the lobule and spreads to the adjacent breast tissues. With passage of time, this lobular carcinoma, which is invasive in nature, spreads to the lymph nodes and to other distal parts. It is more common in older women. HRT increases the incidence of invasive lobular cancer.
4. **Inflammatory Breast Cancer (IBC):** It is a rare and aggressive form of breast cancer. It is aggressive in its truest meaning, as it grows and spreads very quickly. Obesity is a common cause of IBC. Risk of IBC is more in black women. This cancer can affect males as well.
5. **Lobular Carcinoma in Situ (LCIS):** LCIS originates in the lobules, which are the milk producing glands at the end of ducts and it remains confined inside the lobule, rather than spreading to other parts of the body. More than one lobule may be affected. LCIS increases the chances of invasive breast cancer. Truly speaking, LCIS is not a breast cancer, however, it is a warning that breast cancer may be knocking at the door. Since LCIS is not true cancer, it is often called lobular neoplasia (a cluster of abnormal cells). LCIS may be diagnosed before menopause and it hardly occurs in males.

6. **Male Breast Cancer:** Men can have breast cancer, though incidences are only 1% of all breast cancers. Like women, men also have breast tissues, though they do not produce breast-stimulating hormones. As a matter of fact, male breasts are flat and small. Though incidence of male breast cancer is low, chance of survival is low with higher mortality rates (Konduri et al. 2020).

1.5 Molecular Subtypes of Breast Cancer

Different types of breast cancer that have been mentioned above grow and behave differently. Therefore, the need for defining breast cancer at the molecular level is important. Classification at the molecular level considers the type of cancer cells present, their pattern and characteristics. This is important to design a treatment regimen. There are no unique treatments for breast cancer, different types respond to different treatments. The main subtypes of breast cancer are discussed below, and the classification of these subtypes is based on the genes expressed (Yersal and Barutca 2014).

1. **Luminal A:** They tend to grow at a slow pace. Luminal A subtype is hormone-receptor positive, as they are defined by their hormone receptors—oestrogen receptor (ER)-positive and/or progesterone receptor (PR)-positive. They are human epithelial growth factor receptor 2 (HER2) negative. The level of nuclear protein proliferation marker Ki-67, which is a prognostic indicator, is low. This subtype has a good prognosis as the cancer grows slowly.
2. **Luminal B:** In this subtype, cancer cells grow faster and are aggressive in nature. They are hormone-receptor positive (oestrogen receptor and/or progesterone receptor positive), HER2 status may be positive or negative. Proliferative marker Ki-67 is high. In terms of prognosis, it is not very hopeful.
3. **Triple negative:** This type of breast cancer is hormone receptor-negative (oestrogen receptor and progesterone receptor-negative) and HER2 negative. Being devoid of all three receptors of ER, PR, and HER2, this type is called triple negative. Therefore, this type of cancer cannot be treated by hormone. This type of cancer originates in the ducts of the breast and is invasive, and prognosis is not good.
4. **HER2-positive or HER2-enriched:** This type is a fast-growing cancer, with a poor prognosis. Being HER-positive, this cancer can be treated with targeted therapies that target the HER2 protein. This cancer is ER and PR negative, but has many copies of HER2 gene which make HER2 receptors. During proliferation, these receptors guide the cells to divide and grow uncontrollably.
5. **Normal-like:** This has some resemblance to Luminal A, both are ER positive, PR positive, and HER2 negative. Expression of the proliferative marker Ki-67 is low and therefore growth of cancer cells is well controlled. Prognosis of this type is not bad, but not as good as Luminal A type.
6. **Paget's disease of the nipple:** This is a rare form of cancer, where the cancer grows in the nipple and around it. Initially it affects the milk carrying ducts of the

nipple and then advances to the areola which is the dark circle around the nipple. The cancer affected parts become red, scaly with a tingling sensation. Changes in nipple and areola are the first signs of breast cancer, which may be DCIS or invasive cancer. Though this is common in women, men can be affected as well.

7. **Phyllodes tumours of the breast:** The tumour cells grow in a pattern resembling a leaf, hence the name phyllodes, meaning leaflike. These tumours are very fast growing, but they hardly invade outside tissues. Phyllode tumours are mostly benign, but some are cancerous, some may be ambivalent, (borderline) i.e. in between benign and malignant. This type of tumour is rare in men.
8. **Recurrent breast cancer:** When the cancer reappears in the same breast, or on the opposite breast, it is called recurrent breast cancer. Surgery, radiation, and chemotherapy may miss some cancer cells and tests often fail to detect tiny groups of cancer cells. They may start multiplying and grow into a tumour. This comeback of breast cancer may be seen in the site where it was diagnosed, which is called local recurrence. It may be in the lymph nodes or other areas near the site of the original diagnosis, this is termed regional recurrence. It might affect some other distal part of the body like brain, bones, lungs, etc. which is known as distal recurrence. Cancer in the other breast belongs to this category.
9. **Metastatic breast cancer:** When cancer cells migrate from the main tumour and enter the bloodstream or lymphatic system, it is called "metastatic cancer", "advanced cancer", or "stage 4 cancer". Women diagnosed with initial stage breast cancer may develop metastatic disease later on. A metastatic tumour at a distal part of the body (in a different organ) is made up of cells from the breast cancer. Advancement of treatment may control the disease for a number of years.

Different types and subtypes of breast cancer are depicted in Fig. 1.2.

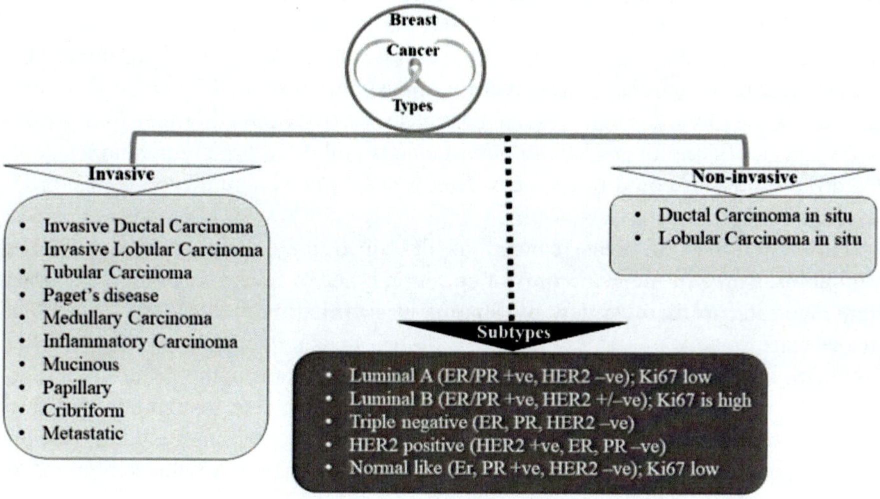

Fig. 1.2 Classification of different types and subtypes of breast cancer

1.6 Signs and Symptoms of Breast Cancer

Signs and symptoms usually indicate the physiological changes in the body. Though mentioned together, there are some differences between signs and symptoms and this difference is based on the observer. If the problem is noticed by the patient, it is called a symptom. If the problem is noticed by medical personnel, it is a sign. If noticed by the patient, as well as a doctor, it is a sign and a symptom. Signs and symptoms aid in identification of medical problems.

In case of a dreadful disease like cancer, signs and symptoms play important roles in control of the disease. Early detection of cancer, before it has progressed to a much-advanced stage is a key to cancer control. Some symptoms may give warnings that it is time to see a doctor and seek medical help. Ignorance and delay cause poor survival, associated problems pile up and cost of treatment often becomes a huge economic burden. Early diagnosis enhances the outcome of the disease by providing care at an early stage of the disease. In order to achieve these, a good knowledge about the disease can guide the patient to the right track, so that cure can be achieved.

Painless lump or thickening of the breast may be a sign of breast cancer. If an abnormal lump in the breast is felt by hand, medical help should be sought without delay. Pain in armpit or breast which does not disappear after the monthly cycle is a point of concern. Alteration in size of the breast, its shape or any changes in appearance are of great concern. Reddening of the breast, dimpling or any change in the skin of the breast like flaking, peeling, or scaling, alteration in the appearance of the nipple, rash around nipple or areola, discharge from the nipple—all these signs should never be ignored. Discharge may sometimes be clear, sometimes bloody or may be of any other colour. Often a nipple may appear sunken or inverted. Of the two, one breast may look different from the other one. Any lump in the breast does not necessarily mean development of cancer. Usually, most of the lumps are non-cancerous; they may be fibroadenomas or cysts, but probability of cancer needs to be ruled out in the first place, and that too by an oncologist. If the cancer has progressed, it may penetrate the skin and ulceration may be noticed. If there is a persistent sore, which may not be painful, one should go for a biopsy after seeing a doctor. As the disease progresses, cancer cells may spread to other parts of the body and spark new symptoms. Usually, breast cancer cells affect the lymph nodes under the armpit, but this could be painless. Breast cancer cells may invade bones, brain, liver, and lungs if metastasis occurs.

Different types of breast cancer (as described previously) may have varied symptoms. Some of the symptoms of common types of breast cancer are discussed here. We refer to the overview by Sharma et al. (Sharma et al. 2010) for detailed discussions.

DCIS, the most common type of breast cancer originates in the ducts and has not spread to other areas. Presence of ductal carcinoma often remains unnoticed. Sometimes, a breast lump may be found, or some discharge containing blood may occur.

Lobular carcinoma which begins in the lobules may have the symptoms of fullness, swelling, and thickening in the area; nipples become flat or inverted.

Invasive carcinoma of the breast may spread from its origin to the surrounding tissues. A lump in the breast or armpit may be noticed, which seems to be attached to the skin. Two breasts seem to be different. Skin over the breast may be thick, red and dimpled, or there may be rash. Sores on the skin may be seen, and the breast may be swollen. Lymph nodes may seem to be joined together and stuck to the skin. Often pain may be felt.

The growth spreads to different organs in metastatic breast cancer. Some of the symptoms are headache, pain in the bone, compromise of brain function, breathing problem, swelling of the abdomen, nausea, anorexia, weight loss, myasthenia or weakness in muscle, double vision, and jaundice like symptoms (yellow skin and eyes).

One of the most aggressive breast cancers, the triple-negative breast cancer, exhibits the same symptoms as other common types. Symptoms are a lump or mass in the breast, redness and painful breast, retracted nipple and may be some discharge from the nipple.

Paget's disease may affect skin of the nipple and areola and may have eczema like symptoms. Skin of the nipple may be red and scaly. Burning or itching sensation may be felt. Discharge from the nipple is seen that discharge may be red or yellow. Flat or inverted nipple is often found.

There is a chance of recurrence of breast cancer within 2 years after the treatment. The risk goes down beyond that period. When breast cancer comes back, swelling or lump under the armpit, above the collarbone or on the chest may occur. Prolonged chest pain may be felt. There may be torpidity (numbness), or pain in the arm or in the shoulder. Another clinical manifestation of relapse of breast cancer is dysphagia or trouble in swallowing.

For metastatic recurrence, symptoms depend on the organ affected. Breast cancer usually metastasizes to the bones, brain, liver, and lungs. Common symptoms are ostealgia or pain in the bone, anorexia and weight loss, severe headache, dry cough, vision problem, disequilibrium or balance problems, a sudden and uncontrolled electrical disturbance in the brain called seizure, and a confused state of mind.

Symptoms of male breast cancer normally go unnoticed because male breast cancer is unknown to many. Male breast cancer is usually detected at an advanced stage. Therefore, it is very important to know the symptoms of male breast cancer. A lump in the male breast or armpit should not be ignored. Like female breast cancer, changes in the skin of breast or nipple, redness, wrinkle, scales, or discharge are some of the symptoms.

1.7 Stages of Breast Cancer

Breast cancer originates in the epithelial lining cells of the ducts or lobules in the glandular tissue and therefore at the initial stage, malignant growth is confined to the duct or lobule. This is called "in situ" and designated as stage 0. With time, stage 0 cancer may progress and spread to the surrounding breast tissues and then to nearby lymph nodes (regional metastasis) and finally to other organs, which is called distant metastasis.

The TNM staging system determines the overall stage of cancer. Staging describes the size of the cancer and the extent of the growth. It is customary to note how big the cancer is, if it has spread to the surrounding tissues, and if it has spread to other organs. Staging is very important, as it guides the treatment modality. TNM staging system refers to tumour, node, and metastasis. T refers to the size of the tumour, N represents if there are any cancer cells in the lymph nodes in the armpits, neck, and inside the chest, M describes if the offending growth has spread to other parts of the body. T which describes the size of tumour may have the values 1, 2, 3, or 4, of these 1 is the lowest one and 4 the largest. Involvement of nodes assigned by N may have the values 0, 1, 2, or 3; 0 means no involvement of lymph nodes and 3, which is the highest means involvement of lots of lymph nodes. M may be 0 or 1; 0 means there is no spread of cancer and 1 means that the cancer has spread. For example, T2 N1 M0 indicates that the cancer is small, which has affected the lymph nodes, but has not spread to any other part. T4 N3 M1 describes a very advanced stage of cancer, i.e. tumour size is big, several lymph nodes are affected and the cancer has spread to distant locations. The second example, being that of a larger size tumour with a nodal spread and metastasis has a worse prognosis. Due to unavailability of immunohistochemical reports, anatomic staging is well accepted (Koh and Kim 2019).

Stage 0 is a pre-cancerous condition or in situ stage, either DCIS or LCIS, where cancer is confined within the duct or lobule of the breast, e.g. Paget's disease.

Stages 1, 2, 3 is the disease within the breast or nearby lymph nodes:

(a) Stage 1: Tumour measures up to 2 cm across, either lymph nodes are not affected, or a few cancer cells are found in the lymph nodes.
(b) Stage 2: The tumour measures 2 cm across and has started to infiltrate to nearby nodes. Also, tumours measuring 2–5 cm across, without any spread to nearby lymph nodes belong to this stage.
(c) Stage 3: In this stage, tumours are up to 5 cm across and have spread to several lymph nodes. Also, tumours measuring more than 5 cm, but very low spread to lymph nodes are designated as stage 3 cancer.
(d) Stage 4 is metastatic cancer, that has poor prognosis. Normally the cancer has spread to distant organs, like bones, brain, liver, or lungs.

Before going into the different types of treatment modality, it is important to have an idea of the diagnosis.

1.8 Diagnosis of Breast Cancer

Diagnosis is the act of identification of a disease from its signs and symptoms. It is of prime importance since the treatment protocol to be followed depends on the diagnosis. Therefore, correct diagnosis is important, which is dependent on a framework of clinical findings. In order to achieve a good outcome, diagnosis has to be accurate, reliable, and prompt.

For cancer, diagnosis is very important, as a quick and accurate diagnosis may save time and helps in treatment and outcome. Diagnosis is done by an expert team, who looks at the cells and tissue samples using a microscope. Often, protein, DNA, and RNA are tested to confirm the presence of cancer. All these help oncologists to formulate the best treatment option. An accurate diagnosis is of much relevance throughout the treatment procedure. Diagnostic tools are important for identifying and locating tumours, staging the disease and determining an appropriate treatment approach. Diagnostic tools help doctors in monitoring the progress of treatment and modifying therapies when needed. There are several tools, which are used to locate the tumour, to determine the stage of the disease, type of cancer, so that a most pertinent treatment modality can be formulated.

For breast cancer, routine screening is needed in order to rule out the presence of any lump, so that the disease can be nipped in the bud. When a person visits a doctor with some symptoms, certain tests are undertaken to confirm diagnosis. Physical examination of breasts by doctor or other health personnel is important; manual examination can help find a lump. Breast self-examination is a screening method to detect any growth in breast, the person needs to sit straight or stand with arms stretching in different positions, e.g. above the head or by their sides (Henderson et al. 2022). However, there is a possibility that a small lump may be missed. Physical examinations are careful manual examinations of breasts by health professionals and doctors. This examination can help find lumps that women may miss with their own self-examinations. Thorough inspection, palpation, and correct documentation are very important. Documentation includes, whether the breasts are symmetrical or asymmetrical; shape of the breast, i.e. if there is any deformity; texture of the breast, i.e. if it is soft, dense presence of ridges; presence of any mass; if the nipple is pink, or brown, if it is inverted or everted, whether there is any discharge from the nipple, if it is dry or scaly. Skin around the breast is also very important to observe, whether it is warm, dry, presence of any sore, oedema, erythema, or redness.

Imaging systems are also employed to scan the breasts. These include mammogram, ultrasound, and Magnetic Resonance Imaging (MRI), which are non-invasive diagnostic methods of imaging used for detection of breast cancer. Mammography uses low energy X-rays to examine breast for diagnosis and preliminary screening. This technique is used for early detection of mammary cancer through microcalcifications. For early detection of breast cancer, mammograms are a very good diagnostic method. Even before the presence of a lump is felt, a mammogram can detect it. Mammograms may be used for two purposes, screening and diagnostic. A screening mammogram enables to detect any signs of breast cancer when there are no symptoms. X-ray pictures of both the breasts are taken from two separate angles. When there is a suspicion of breast cancer, diagnostic mammogram is done.

Ultrasound imaging of the breast employs high-frequency sonic waves that produce pictures of the internal structures of the breast. Breast lumps, or other abnormality, detected during physical examination of the breast can be diagnosed.

This is a non-invasive diagnostic tool. Through ultrasound, it can be known if the tumour is solid or it contains liquid.

MRI is a non-invasive test that uses a powerful magnetic field. Radio frequency pulses produce images of the breast, and a computing system aids to get a detailed 3-dimensional image of the breast tissue. Cross-sectional images can also be seen.

CT scans, bone scans, or PET scans are employed to find the spread of breast cancer, if any. 3D mammography or breast tomosynthesis is performed often, by which breast tissues can be seen clearly in three dimensions. It is very helpful for dense breasts.

For women who are younger than 35, it is suggested to have ultrasound of the breast. Mammogram is not very effective for younger women with dense breasts.

With the imaging tools mentioned above, doctors may have an idea that there is a possibility of breast cancer. This suspicion needs to be confirmed for further action. Confirmation of breast cancer can be done with the help of biopsy. This is a small procedure, where a tissue is taken from the suspected part of the breast. The tissue taken is examined by a pathologist to see the presence of cancer cells. If cancer cells are seen, the characteristics of the cells are studied in detail. Type of cancer can be determined also through biopsy, it can be found if the cancer is hormone sensitive or hormone insensitive, which is important to set up a treatment regimen.

Biopsies may be taken in a number of ways, needle aspiration, needle biopsy, and vacuum-assisted biopsy. In needle aspiration (also called fine needle aspiration biopsy), a thin small needle is inserted to take out cells from breast, without removing any tissue. From a small fluid filled lump, which is normally a benign cyst, cells are taken out for examination. In needle biopsy (also called core needle biopsy), a sample of tissue is taken from a lump in the breast using a large wider needle. This is a diagnostic technique followed to confirm cancerous growth, if it is invasive or not, which has been indicated by physical examination or imaging. Also, the aim is to find the hormone-receptor status and also HER2neu and Ki-67 status (ER, PR, HER2, Ki-67). For guided needle biopsy, ultrasound, X-ray, or MRI may be used to get a precise and accurate diagnosis. Surgical biopsy is also performed, which can remove a substantial part of the tumour. There is another method, called image-guided biopsy, a needle is directed to the site of the growth by imaging techniques, e.g. mammography, ultrasound, or MRI. A titanium clip is put into the breast to mark the exact position of the biopsy. This helps to know the exact position of the growth in case surgery is needed after diagnosis. Another method of biopsy is sentinel lymph node biopsy. If the cancer has spread to lymph nodes, the first one is known as sentinel lymph node. This type of biopsy helps to find if the growth has affected the lymph nodes. There is one more type of biopsy, called vacuum-assisted biopsy or Mammotome. In this procedure, a suction tube is attached to the needle, that helps to get the sample, and at the same time any bleeding from that area may be cleared.

There are a few more diagnostics tests. These are (a) nuclear medicine tests or radionuclide imaging (tracer or radioactive molecules are injected into the blood stream, which go to the cancer cells and are detected by a camera); (b) Molecular Breast Imaging or MBI or **scintimammography** or BSGI, i.e. **breast-specific**

gamma imaging, where a tracer called techmetium-99 m is injected into the blood; (c) PET scan or positron **emission tomography (PET) scan**, a radioactive sugar (tracer) is injected into the blood. This is used to diagnose spread of cancer to other parts; (d) PEM or **Positron Emission Mammography which is a combination of PET and mammography;** (e) CEM or CESM, i.e. **Contrast-Enhanced Spectral Mammography (CESM) which** is a relatively recent test in which a contrast dye containing iodine is injected into the blood a few minutes before two sets of mammograms; (f) Elastography which is done as part of an ultrasound examination. The rationale is that breast tumours tend to be firmer and stiffer than the surrounding breast tissue; (g) Optical imaging test where the light that passes through the tumour is measured; (h) Electrical Impedance Tomography or EIT, which is based on the fact that breast cancer cells conduct electricity differently from normal cells. He et al. (He et al. 2020) review the diagnostic tests comprehensively.

1.9 Treatment of Breast Cancer

Treatments consist of management and care of patients to combat a disease or disorder. Following diagnosis of any disease, the next step is to think about a therapy or treatment which will lead to remediation of the disease. Therapy may have some pros and cons, and there may be a number of ways a therapy can be decided. All the available therapies may not be equally effective, sometimes therapy may have some serious adverse effects. Therapies or treatment protocols are decided considering the details of the disease as understood from diagnosis.

For cancer, a prior knowledge of the treatment to be given to a patient is very important, because these treatments are not conventional and there are lots of side effects. For breast cancer, treatment depends on a number of factors. We have learnt that there are several types of breast cancer, each of which needs a different treatment. The important aspects to consider for treatment of breast cancer are, size of the growth, molecular subtypes, whether hormones are involved, and whether metastasis has occurred. Each treatment option has its merits and demerits; therefore, it is of enormous importance to know the benefits of treatment method against the possible risk factors and side effects that may be caused.

Surgery and radiation therapy: Surgery and radiation therapy are local, as surgery removes the offending growth and radiation treats the tumour without affecting other parts of the body. Breast surgery may be of different types, which are situation dependant. Surgery removes the cancer as far as possible along with some surrounding normal tissue, as in breast-conservation surgery (also known as lumpectomy, partial mastectomy). There is another type, called mastectomy, i.e. complete removal of the breast, along with nearby tissues. After removal of the growth, shape of the breast is restored, which is called breast reconstruction surgery. When it is diagnosed that the cancer has spread to lymph nodes in the armpit, sentinel lymph node biopsy or auxiliary lymph node dissection is done. Surgery also relieves symptoms of advanced cancer. While spreading, cancer cells first target the lymph nodes and these are called sentinel lymph nodes. Presence of cancer cells in the sentinel nodes warrants further surgery to get rid of them.

High energy rays can kill cancer cells, which is the rationale for radiation therapy. Often radiation therapy is given in addition to other treatment modalities. Sometimes breast conserving surgery is followed by radiation treatment for preventing the recurrence of cancer. Radiation therapy is also given after mastectomy if the cancer size is somewhat bigger. Also, if cancer has metastasised to the bones, brain, or spinal cord, radiation therapy is given. Different types of radiation therapy used are, External Beam Radiation Therapy (EBRT) and brachytherapy. There are certain side effects of radiotherapy. Darkening of the skin and irritation may occur, which may lead to sore and redness. Fatigue is another problem of radiation therapy. Often there is fluid accumulation in the arm.

Systemic therapies: Drugs may reach cancer cells anywhere in the body. These therapies may be given orally, intramuscular, or put directly into the blood stream. Choice of treatment depends on the type of cancer. These are chemotherapy, hormone therapy, targeted therapy, and immunotherapy. Chemotherapy consists of anti-cancer drugs that may be given intravenously or orally. If the cancer cells spread to the spinal cord, chemo drugs are delivered directly into the affected part, called intrathecal chemotherapy. Chemotherapy also has several side effects; these are loss of appetite, tiredness or fatigue, hair loss, nausea, mouth sore, etc.

Hormone (oestrogen and progesterone) dependent breast cancer that are affected by hormones is so treated that functioning of hormones is blocked. These are called hormone or endocrine therapy. In targeted therapy, drugs are targeted to the proteins that help in abnormal proliferation of cancer cells. This therapy too has its side effects such as diarrhoea, nausea, headache, shivering, cough, skin rash are among them.

Another treatment modality is immunotherapy, administration of which boosts the immune system and destroys cancer cells. Application of immunotherapy has been found to be quite promising (Williams et al. 2017). Treatment plans are also dependent on the subtypes of breast cancer. For HR-positive luminal A, surgery is the choice of treatment, followed by hormonal therapies, when the cancerous growth is hormone dependent. Cancers belonging to Luminal B are treated with chemotherapy and therapy that targets HER2. For triple-negative breast cancer, chemotherapy is the main treatment option. Initially chemo drugs work well, but there is a possibility of recurrence. For stage 1 to 3 TNBC tumour, if the tumour size is not too big, surgery is done, followed by breast conserving surgery or mastectomy. For large tumours, generally surgery followed by radiation is practiced. Adjuvant chemotherapy, i.e. chemotherapy after surgery is often given to eliminate the chances of recurrence of cancer. Sometimes, chemotherapy is given before surgery, which is called neoadjuvant chemotherapy. This is done with the aim to shrink the tumour before surgery. For stage 4 TNBC, chemotherapy is given when cancer has spread to other parts. For advanced TNBC, immunotherapy and chemotherapy are given (Wahba and El-Hadaad 2015).

For HER2-positive cancerous growth, monoclonal antibodies are generally used, alone or in combination with chemotherapy. Adjuvant chemotherapy may be given for patients with infiltrating breast cancer, where the first line of treatment is surgery.

Cancer patients need psychological support. The thought of suffering from breast cancer is traumatic, there are a number of distressing emotions, which may cause depression, anxiety, and dilemma or confusion. There are certain practical difficulties and stress. Removal of a part of the breast or the whole of it is very stressful and traumatic. Family relationship is often strained, and financial burden is another point of concern along with physical illness. Talking to someone who has gone through this situation may be of help. Sharing the feelings with others is very important, rather than keeping the thoughts within oneself. Professional help may be important to overcome these unsettling feelings.

1.10 Conclusion

We have given an overview of breast cancer in this chapter. Breast cancer is a complex disease, and its treatment depends on accurate diagnosis of the disease. The research in breast cancer is primarily targeted towards the understanding of the disease both from physiological aspects like the types of tumours and their spread, as well as the understanding of the molecular, genetic, and epigenetic mechanisms underlying the disease. A thorough understanding of these mechanisms can aid in developing new drugs and treatment options for breast cancer. We have reviewed the epidemiology, types, molecular mechanisms, diagnosis, treatment, and signs and symptoms of breast cancer in this chapter. We discuss the details of these aspects in the following chapters, namely the genetic landscape of breast cancer, the epigenetic mechanisms of breast cancer and therapy, and control and prevention of the disease.

References

Al Jishi T, Sergi C (2017) Current perspective of diethylstilbestrol (DES) exposure in mothers and offspring. Reprod Toxicol 71:71–77. https://doi.org/10.1016/j.reprotox.2017.04.009. Epub 2017 Apr 28

Anonymous (1976) Exposure in utero to diethylstilbestrol and related synthetic hormones. Association with vaginal and cervical cancers and other abnormalities. JAMA 236(10):1107–1109. https://doi.org/10.1001/jama.1976.03270110011002

Begg CB, Ostrovnaya I, Geyer FC, Papanastasiou AD, Ng CKY, Sakr RA, Bernstein JL, Burke KA, King TA, Piscuoglio S, Mauguen A, Orlow I, Weigelt B, Seshan VE, Morrow M, Reis-Filho JS (2018) Contralateral breast cancers: independent cancers or metastases? Int J Cancer 142(2):347–356. https://doi.org/10.1002/ijc.31051. Epub 2017 Sep 28

Benz CC (2008) Impact of aging on the biology of breast cancer. Crit Rev Oncol Hematol 66(1): 65–74. https://doi.org/10.1016/j.critrevonc.2007.09.001

Beral V, Million Women Study Collaborators (2003) Breast cancer and hormone-replacement therapy in the million women study. Lancet 362(9382):419–427. https://doi.org/10.1016/s0140-6736(03)14065-2. Erratum in: Lancet. 2003; 362 (9390): 1160

Boyd NF, Martin LJ, Bronskill M, Yaffe MJ, Duric N, Minkin S (2010) Breast tissue composition and susceptibility to breast cancer. J Natl Cancer Inst 102(16):1224–1237. https://doi.org/10.1093/jnci/djq239

Cao Y, Karin M (2003) NF-kappaB in mammary gland development and breast cancer. J Mammary Gland Biol Neoplasia 8(2):215–223. https://doi.org/10.1023/a:1025905008934

Chlebowski RT, Chen Z, Anderson GL, Rohan T, Aragaki A, Lane D, Dolan NC, Paskett ED, McTiernan A, Hubbell FA, Adams-Campbell LL, Prentice R (2005) Ethnicity and breast cancer: factors influencing differences in incidence and outcome. J Natl Cancer Inst 97(6):439–448. https://doi.org/10.1093/jnci/dji064

Choi KC, Kang SK, Tai CJ, Auersperg N, Leung PC (2001) Estradiol up-regulates antiapoptotic Bcl-2 messenger ribonucleic acid and protein in tumorigenic ovarian surface epithelium cells. Endocrinology 142(6):2351–2360. https://doi.org/10.1210/endo.142.6.8144

Chung SS, Giehl N, Wu Y, Vadgama JV (2014) STAT3 activation in HER2-overexpressing breast cancer promotes epithelial-mesenchymal transition and cancer stem cell traits. Int J Oncol 44(2): 403–411. https://doi.org/10.3892/ijo.2013.2195. Epub 2013 Nov 29

Collaborative Group on Hormonal Factors in Breast Cancer (2012) Menarche, menopause, and breast cancer risk: individual participant meta-analysis, including 118 964 women with breast cancer from 117 epidemiological studies. Lancet Oncol 13(11):1141–1151. https://doi.org/10.1016/S1470-2045(12)70425-4

Eidemüller M, Holmberg E, Lundell M, Karlsson P (2021) Evidence for increased susceptibility to breast cancer from exposure to ionizing radiation due to a familial history of breast cancer: results from the Swedish hemangioma cohort. Am J Epidemiol 190(1):76–84. https://doi.org/10.1093/aje/kwaa163

Ferlay J, Ervik M, Lam F, Colombet M, Mery L, Piñeros M, Znaor A, Soerjomataram I, Bray F (2020) Global cancer observatory: cancer today. International Agency for Research on Cancer, Lyon. https://gco.iarc.fr/today

Filipits M, Nielsen TO, Rudas M, Greil R, Stöger H, Jakesz R, Bago-Horvath Z, Dietze O, Regitnig P, Gruber-Rossipal C, Müller-Holzner E, Singer CF, Mlineritsch B, Dubsky P, Bauernhofer T, Hubalek M, Knauer M, Trapl H, Fesl C, Schaper C, Ferree S, Liu S, Cowens JW, Gnant M, Austrian Breast and Colorectal Cancer Study Group (2014) The PAM50 risk-of-recurrence score predicts risk for late distant recurrence after endocrine therapy in postmenopausal women with endocrine-responsive early breast cancer. Clin Cancer Res 20(5): 1298–1305. https://doi.org/10.1158/1078-0432.ccr-13-1845

Ghosh S, Karin M (2002) Missing pieces in the NF-kappaB puzzle. Cell 109 Suppl:S81–S96. https://doi.org/10.1016/s0092-8674(02)00703-1

Gong Y, Chippada-Venkata UD, Oh WK (2014) Roles of matrix metalloproteinases and their natural inhibitors in prostate cancer progression. Cancers (Basel) 6(3):1298–1327. https://doi.org/10.3390/cancers6031298

Hanahan D, Weinberg RA (2011) Hallmarks of cancer: the next generation. Cell 144(5):646–674. https://doi.org/10.1016/j.cell.2011.02.013

Hanzlikova H, Gittens W, Krejcikova K, Zeng Z, Caldecott KW (2017) Overlapping roles for PARP1 and PARP2 in the recruitment of endogenous XRCC1 and PNKP into oxidized chromatin. Nucleic Acids Res 45(5):2546–2557. https://doi.org/10.1093/nar/gkw1246

He Z, Chen Z, Tan M, Elingarami S, Liu Y, Li T, Deng Y, He N, Li S, Fu J, Li W (2020) A review on methods for diagnosis of breast cancer cells and tissues. Cell Prolif 53(7):e12822. https://doi.org/10.1111/cpr.12822

Henderson JA, Duffee D, Ferguson T. Breast examination techniques. In Stat Pearls. Treasure Island (FL): Stat Pearls Publishing; 2022. https://www.ncbi.nlm.nih.gov/books/NBK459179/

Henrich LM, Smith JA, Kitt D, Errington TM, Nguyen B, Traish AM, Lannigan DA (2003) Extracellular signal-regulated kinase 7, a regulator of hormone-dependent estrogen receptor destruction. Mol Cell Biol 23(17):5979–5988. https://doi.org/10.1128/MCB.23.17.5979-5988.2003

Husby A, Wohlfahrt J, Øyen N, Melbye M (2018) Pregnancy duration and breast cancer risk. Nat Commun 9(1):4255. https://doi.org/10.1038/s41467-018-06748-3

Inwald EC, Klinkhammer-Schalke M, Hofstädter F, Zeman F, Koller M, Gerstenhauer M, Ortmann O (2013) Ki-67 is a prognostic parameter in breast cancer patients: results of a large population-based cohort of a cancer registry. Breast Cancer Res Treat 139(2):539–552. https://doi.org/10.1007/s10549-013-2560-8. Epub 2013 May 16

Khan S, Lopez-Dee Z, Kumar R, Ling J (2013) Activation of NFκB is a novel mechanism of pro-survival activity of glucocorticoids in breast cancer cells. Cancer Lett 337(1):90–95. https://doi.org/10.1016/j.canlet.2013.05.020. Epub 2013 May 18

Koh J, Kim MJ (2019) Introduction of a new staging system of breast cancer for radiologists: an emphasis on the prognostic stage. Korean J Radiol 20(1):69–82. https://doi.org/10.3348/kjr.2018.0231. Epub 2018 Dec 27. Erratum in: Korean J Radiol. 2022; 23(5): 570

Konduri S, Singh M, Bobustuc G, Rovin R, Kassam A (2020) Epidemiology of male breast cancer. Breast 54:8–14. https://doi.org/10.1016/j.breast.2020.08.010. Epub 2020 Aug 22

Łukasiewicz S, Czeczelewski M, Forma A, Baj J, Sitarz R, Stanisławek A (2021) Breast cancer–epidemiology, risk factors, classification, prognostic markers, and current treatment strategies–an updated review. Cancers (Basel) 13:4287. https://doi.org/10.3390/cancers13174287

Lynch BM, Neilson HK, Friedenreich CM (2011) Physical activity and breast cancer prevention. Recent Results Cancer Res 186:13–42. https://doi.org/10.1007/978-3-642-04231-7_2

Mohibi S, Mirza S, Band H, Band V (2011) Mouse models of estrogen receptor-positive breast cancer. J Carcinog 10:35. https://doi.org/10.4103/1477-3163.91116. Epub 2011 Dec 22

Mørch LS, Skovlund CW, Hannaford PC, Iversen L, Fielding S, Lidegaard Ø (2017) Contemporary hormonal contraception and the risk of breast cancer. N Engl J Med 377(23):2228–2239. https://doi.org/10.1056/NEJMoa1700732

Narod SA (2011 Aug 2) Hormone replacement therapy and the risk of breast cancer. Nat Rev Clin Oncol 8(11):669–676. https://doi.org/10.1038/nrclinonc.2011.110

Narod SA, Tung N, Lubinski J, Huzarski T, Robson M, Lynch HT, Neuhausen SL, Ghadirian P, Kim-Sing C, Sun P, Foulkes WD, Hereditary Breast Cancer Clinical Study Group (2014) A prior diagnosis of breast cancer is a risk factor for breast cancer in BRCA1 and BRCA2 carriers. Curr Oncol 21(2):64–68. https://doi.org/10.3747/co.21.1656

Palmer JR, Wise LA, Hatch EE, Troisi R, Titus-Ernstoff L, Strohsnitter W, Kaufman R, Herbst AL, Noller KL, Hyer M, Hoover RN (2006) Prenatal diethylstilbestrol exposure and risk of breast cancer. Cancer Epidemiol Biomark Prev 15(8):1509–1514. https://doi.org/10.1158/1055-9965.EPI-06-0109

Petrucelli N, Daly MB, Pal T (1998) BRCA1- and BRCA2-associated hereditary breast and ovarian cancer. In: Adam MP, Ardinger HH, Pagon RA et al (eds) GeneReviews®. University of Washington, Seattle (WA). https://www.ncbi.nlm.nih.gov/books/NBK1247/

Ronckers CM, Erdmann CA, Land CE (2005) Radiation and breast cancer: a review of current evidence. Breast Cancer Res 7(1):21–32. https://doi.org/10.1186/bcr970

Roy M, Datta A (2019) Cancer genetics and therapeutics: focus on phytochemicals, 1st edn. Springer, Singapore. ISBN: 978-981-13-9471-3

Sharma R (2021) Global, regional, national burden of breast cancer in 185 countries: evidence from GLOBOCAN 2018. Breast Cancer Res Treat 187(2):557–567. https://doi.org/10.1007/s10549-020-06083-6. Epub 2021 Jan 30

Sharma GN, Dave R, Sanadya J, Sharma P, Sharma KK (2010) Various types and management of breast cancer: an overview. J Adv Pharm Technol Res 1(2):109–126

Sirotkovic-Skerlev M, Kuli A, Bradi L, Cacev T (2012) Protumor effects of proinflammatory mediators in breast cancer. Period Biol 114(4):489–496

Sun YS, Zhao Z, Yang ZN, Xu F, Lu HJ, Zhu ZY, Shi W, Jiang J, Yao PP, Zhu HP (2017) Risk factors and preventions of breast cancer. Int J Biol Sci 13(11):1387–1397. https://doi.org/10.7150/ijbs.21635

Sung H, Ferlay J, Siegel RL, Laversanne M, Soerjomataram I, Jemal A, Bray F (2021) Global cancer statistics 2020: GLOBOCAN estimates of incidence and mortality worldwide for

36 cancers in 185 countries. CA Cancer J Clin 71(3):209–249. https://doi.org/10.3322/caac.21660. Epub 2021 Feb 4

Turk AA, Wisinski KB (2018) PARP inhibitors in breast cancer: bringing synthetic lethality to the bedside. Cancer 124(12):2498–2506. https://doi.org/10.1002/cncr.31307. Epub 2018 Apr 16

Turnbull C, Rahman N (2008) Genetic predisposition to breast cancer: past, present, and future. Annu Rev Genomics Hum Genet 9:321–345. https://doi.org/10.1146/annurev.genom.9.081307.164339

UICC (2020) GLOBOCAN 2020: new global cancer data. Data source: GLOBOCAN 2020 graph production: IARC (http://gco,iarc,fr/today). https://www.uicc.org/news/globocan-2020-new-global-cancer-data

Veuger SJ, Hunter JE, Durkacz BW (2009) Ionizing radiation-induced NF-kappaB activation requires PARP-1 function to confer radioresistance. Oncogene 28(6):832–842. https://doi.org/10.1038/onc.2008.439. Epub 2008 Dec 8

Wahba HA, El-Hadaad HA (2015) Current approaches in treatment of triple-negative breast cancer. Cancer Biol Med 12(2):106–116. https://doi.org/10.7497/j.issn.2095-3941.2015.0030

Welcsh PL, King MC (2001) BRCA1 and BRCA2 and the genetics of breast and ovarian cancer. Hum Mol Genet 10(7):705–713. https://doi.org/10.1093/hmg/10.7.705

Williams AD, Payne KK, Posey AD Jr, Hill C, Conejo-Garcia J, June CH, Tchou J (2017) Immunotherapy for breast cancer: current and future strategies. Curr Surg Rep 5:31. https://doi.org/10.1007/s40137-017-0194-1

Yersal O, Barutca S (2014) Biological subtypes of breast cancer: prognostic and therapeutic implications. World J Clin Oncol 5(3):412–424. https://doi.org/10.5306/wjco.v5.i3.412

Genomic Landscape of Breast Cancer

2

2.1 Introduction

Cancer is primarily a disease that is driven by aberrations in gene expressions. It is important to know which genes are mutated and how mutated genes affect the normal functioning of cells. This understanding is referred to as the genomic landscape of a cancer. This landscape is different for different cancers and varies even within different types of breast cancer. High-throughput new generation sequencing technologies in the last decade have improved our knowledge of mutations in tumor cells. Though we have to remember that the sequencing results are not always very accurate. The first problem is the heterogeneity of tumor cells even within the same tumor and across tumors from different patients. This results in contradictory conclusions from different studies, as we will see in this chapter. There are also technological difficulties, for example, large-scale genome rearrangements occur in tumor cells due to gene fusions and chromosome translocations, resulting in inaccuracies in the sequencing results. Hence it is important to keep in mind that our understanding of the genomic landscapes of different types of breast cancer or other cancers is still incomplete and is a very active area of research.

Mutations in genes occur in all cells during cell division, however, DNA repair mechanisms usually correct these mutations. Neoplastic cells on the other hand suffer from runaway mutations in important genes, many of which are tumor suppressor genes, their role is to maintain the normal metabolic functioning of cells and also maintain the expression levels of oncogenes or cancer-causing genes at a normal level. Even a few mutations in different genes may give a cell growth advantage over normal cells, however malignancy or unrestricted growth requires many coordinated mutations in a cell. It is well known that the number of cells in a tumor with a large number of mutations is far higher than the number of cells with a low number of mutations. This clearly indicates that a large number of cells with many mutations are required for a tumor to become malignant. Also, the number of mutations in genes varies at different stages of a tumor. This is another difficulty in

M. Roy et al., *Genetics and Epigenetics of Breast Cancer*,
https://doi.org/10.1007/978-981-19-9925-3_2

comparing the results from different studies, as different studies may take samples from tumors at different stages.

Another difficulty in determining the genomic landscape of a cancer is differentiating between "passenger" and "driver" mutations. Usually, a large number of mutated genes are detected in each study, but many of these mutations may not have direct links to malignancy, these are called passenger mutations. On the other hand, some mutations are key to malignancy and are called driver mutations. A common way of determining driver mutations is to examine the degree of mutations in a gene, e.g., single nucleotide variations, copy number variations, genome rearrangements, etc. Usually highly mutated genes are identified as driver genes for malignancy.

We will review the current understanding of the genomic landscapes of ER+, ER−, HER2-positive, basal-like and triple-negative breast cancer, ductal carcinoma in situ, and invasive lobular carcinoma in this chapter. However, we start our review by discussing two tumor suppressor genes BRCA1 and BRCA2, the first genes that were associated with hereditary breast cancer susceptibility.

2.2 BRCA1 and BRCA2

BRCA1 and BRCA2 are two of the most well known tumor suppressor genes, and mutations of these two genes can increase the chance of breast cancer significantly. The coding regions of these two genes do not show any homology between the two genes. Hereditary breast and ovarian cancer (HBOC) syndrome results if even one copy of one or both of these genes is mutated in the germline. The mutations may be inherited from either parent. Even a single mutated copy of BRCA1 or BRCA2 can increase the chances of breast, ovarian, pancreatic, laryngeal, stomach, and prostate cancers. The chance of developing breast cancer is the highest, about 5–7% of all breast cancers are due to HBOC (Roy et al. 2011). The lifetime chance of developing breast cancer due to HBOC is 50–80% higher and that of ovarian cancer is 30–50% higher (Roy et al. 2011). BRCA1 associated HBOC results predominantly in estrogen receptor (ER) negative breast cancer, however BRCA2 associated breast cancer does not have any such bias. Understanding the roles of BRCA1 and BRCA2 in carcinogenesis is an important research area. Many of the aspects related to BRCA1 and BRCA2 are not well understood. We will first discuss these two genes in detail before discussing the broader genetic landscape of breast cancer.

Knudson's "two-hit" hypothesis (Chial 2008; Knudson 2001) is widely assumed to be a guiding principle in cancer research. Since each gene has two copies, both the copies must be inactive for a tumor suppressor gene to be completely inactive (loss of heterogeneity or LOH), a copy of the gene may become inactive either due to mutation, or due to post-translational epigenetic silencing. And this has been found to be the case in many cancers. BRCA1 is a tumor suppressor gene for both ovarian and breast cancers, and both copies of BRCA1 are mutated in ovarian cancer. However, BRCA1 is known to initiate carcinogenesis in breast cancer even when just one copy is mutated (Meric-Bernstam 2007) and there is no evidence of epigenetic modulation of the other copy (Semmler et al. 2019). This makes it difficult to understand the exact role of BRCA1 in carcinogenesis in breast cancer.

As BRCA1 is a DNA repair gene, its inactivation should affect all cells, not just the epithelial cells where breast carcinogenesis occurs. BRCA1 initiated repair mechanisms are particularly important for the most severe DNA damage, DNA double-strand break (DSB). However, there is almost no effect of BRCA1 inactivation in any other cell. This raises the question if other proteins are responsible for DSB repair in other cell types, why the same proteins are inactive in epithelial cells of breast cancer patients. Furthermore, such proteins, if they exist, must have low expression levels in breast epithelial cells (BEC). However, there is very little progress in the research for finding such tissue-specific expressions of proteins that can compensate for the haplo-deficiency of BRAC1 in non-breast tissues. Figure 2.1 outlines the functions of BRCA1 and BRCA2 genes.

As such, BRCA1 forms complexes with many other proteins, and elucidation of the roles of BRCA1 in many of these complexes is a major thrust in breast cancer research (Ohta et al. 2011; Savage and Harkin 2015; Mullan et al. 2006). BRCA1 is a tumor suppressor gene due to its role in DNA repair mechanisms, in particular repairing DSB and also in stalled replication forks (SRFs) (Semmler et al. 2019; Roy et al. 2011). SRFs delay the cell cycle either before it starts or during the replication process. This is necessary for preventing the damaged DNA to be propagated to daughter cells. SRF allows enough time for the DNA damage repair (DDR) mechanisms to be activated. Both BRCA1 and BRCA2 are DNA repair genes and act at different stages of the DNA repair process. Both proteins encoded by these two genes participate in homologous recombination (HR), the most reliable pathway for DSB repair. Most DSBs occur during replication, and HR-associated repair mechanism uses the sister chromatid that is undamaged for DSB repair with high fidelity (Roy et al. 2011).

BRCA1 is an important protein in many protein complexes that participate in DSB and SRF mechanisms. The BRCA1 C-terminus domain is an important motif that is conserved in many proteins that participate in DSB repair. BRCA1 facilitates forming a complex of proteins that is very important for DSB recognition and repair. Proteins in this complex include BRCA1A complex subunit, CNC homolog 1, BTB domain, and retinoblastoma binding protein 8 (Savage and Harkin 2015). DSB repair through HR requires repositioning of the p53-binding protein 1, a tumor suppressor, from the sites of DSB. This in turn requires the ubiquitination of histone H2A. The complex of BRCA1 and BRCA1 associated ring domain protein (BARD1) participates in this process as an E3 ubiquitin ligase that can transfer ubiquitin molecules to specific substrates (Densham and Morris 2017). Recruiting the RAD51 recombinase to the DSB site is another important step in DSB repair process through homologous recombination. A complex of BRCA1, BRAC2, and PALB2 (partner and localizer of BRCA2) is responsible for this process (Roy et al. 2011).

Though BRCA1 is responsible for signalling for DNA damage, homologous recombination (HR), and other DNA damage repair processes like non-homologous end joining (Bau et al. 2006) and single strand annealing (Anantha

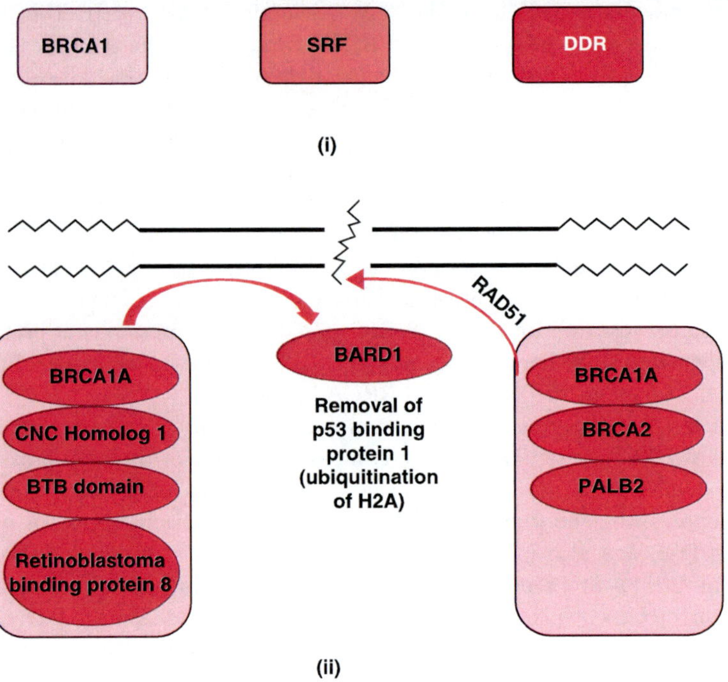

Fig. 2.1 Functions of BRCA1 and BRCA2 genes. (**a**) BRCA1 participates in stalled replication fork (SRF) and DNA damage repair (DDR); (**b**) BRCA1 and BRCA2 participate in DNA damage repair

et al. 2017), BRCA2 almost exclusively participates in HR repair of DSB according to the current understanding. The only known connection between BRCA1 and BRCA2 in this process is the PALB2 protein that forms a complex with BRCA1 and BRCA2 in the HR pathway, which is very important for maintaining the integrity of the genome during its replication. Both PALB2 and BRCA1 have coiled-coil domains. The coiled-coil domain of BRCA1 interacts with the N-terminal coiled-coil domain of PALB2. The N-terminus of BRCA2 interacts with the C-terminus of BRCA2 (Oliver et al. 2009; Zhang et al. 2009). BRCA1 acts as an upstream protein in the HR repair of DSB, as the BRCA1-PALB2 complex forms first, and this complex is important for recruiting both BRCA2 and RAD51 at the site of the DNA damage. The actions of BRCA1 and BRCA2 in repairing DSB are shown in Fig. 2.1.

2.3 Genomic Landscape of ER+ and ER− Breast Cancers

Though BRCA1, BRCA2, and associated genes like PALB2 are some of the earliest genes that were implicated in breast cancer, recent advances in genome wide sequencing studies have identified many more genes related to breast cancer.

However, the genomic landscape varies between patients and even between different cells within the same tumor. Hence research in identifying so-called driver genes, the genes that drive carcinogenesis, is an ongoing effort worldwide. Furthermore, the genomic landscape varies between ER+ and ER− breast cancers significantly. We will base our review on the paper by Stephens et al. (2012) published in 2012 and also review some of the important papers that have been published more recently on the genomic landscape of breast cancer.

ER+ cancer cells get signal from the estrogen hormone, whereas ER− cancer cells do not get this signal. These two types of cancers are about 80% ER+ and 20% ER−. Stephens et al. (2012) studied the genomic landscape of 79 ER+ and 21 ER− breast cancer patients by sequencing the exons of pre-mRNAs of over 21,000 protein-coding genes and over 1600 miRNAs. It is important to study miRNAs as these small non-coding RNAs play important roles in post-transcriptional regulation of mRNAs, by preventing their translation through binding at 3'UTR. We will review the epigenetic roles of miRNA in Chap. 3. Stephens et al. (2012) studied the somatic copy number changes and mutations in exons, as these two can connect particular genes to carcinogenesis. They found strong correlations between the age of the patient and the histological grade with mutations in exons.

This study resulted in confirming several previously known driver genes and identifying some new driver genes. Previously confirmed driver genes included BRCA1, BRCA2, PIK3CA, TP53, AKT1, PTEN, RB1, GATA3, and CDH1; previously suspected driver genes included KRAS, MLL2, MAP2K4, MLL3, ARID2, ARID1A, APC, ASXL1, SETD2, NF1, SF3B1, STK11, BAP1, and SMAD4. They detected several novel driver genes including NCOR1, AKT2, CASP8, ARID1B, MAP3K1, SMARCD1, MAP3K13, TBX3, and CDKN1B. References (NCBI n.d.; Spatz and US National Library of Medicine 2004) are excellent resources for understanding the functions of specific genes, and we explain the roles some of the genes mentioned above play in normal cells (we have reviewed the roles of BRCA1 and BRCA2 previously) and how their mutations may initiate carcinogenesis.

TP53 is perhaps the most important tumor suppressor gene and participates in many important genetic pathways related to cancer. The p53 protein is transcribed from the TP53 gene and among many other roles, it can initiate the apoptosis pathway in response to irreparable DNA damage [(Roy and Datta 2019), Chap. 1]. Mutation in the TP53 gene may result in the loss of ability to trigger the apoptosis pathway, and that in turn may result in uncontrolled proliferation of cells with damaged and mutated genes, a sign of carcinogenesis.

Post-translational modulation of protein functions through phosphorylation is important for controlling the functions of many proteins. Protein kinases are catalysts that attach a phosphate group to a protein to deactivate it. Phosphorylation can be temporary, when other enzymes remove the phosphate group, or permanent for the lifetime of a protein. Phosphorylation may be hindered if protein kinase genes are mutated, and other genes can be overexpressed. The p110α protein is transcribed from the PIK3CA gene and is a catalytic subunit of PI3K (phosphatidylinositol 3-kinase). PI3K is a key member of the PI3K/AKT/mTOR intracellular signalling pathway [(Roy and Datta 2019), Chap. 2], a key pathway for cell cycle regulation, and disruption of which has been implicated as a key driver for several cancers. PTEN is a tumor suppressor gene that transcribes a protein whose role is to remove phosphate groups from proteins after they have been phosphorylated (Chen et al. 2018). Hence PTEN plays a very important role in post-translational modification of many proteins.

Chromosomes are packed inside the nucleus by winding around histone proteins, forming chromatins. We will discuss chromatin remodelling as an epigenetic modulation in Chap. 3. Transcription factors, proteins that initiate the transcription of a gene into pre-mRNA, cannot get access to the genes in many situations. Several proteins help in chromatin remodelling, i.e., loosening of the chromatin to give transcription factors access to certain genes. The driver genes AID2, ARID1A, and ASXL1 transcribe chromatin remodelling proteins. Note that epigenetic modulations like histone methylation and acetylation also act as chromatin remodelling processes (Phillips and Shaw 2008). We will discuss these processes in Chap. 3. SETD2 is a tumor suppressor gene that transcribes a histone methyltransferase protein.

The protein E-cadherin (epithelial cadherin) is transcribed from the CDH1 gene. E-cadherin has a central role in maintaining the phenotype of the extracellular matrix, i.e., to ensure that cells stick to each other in the extracellular matrix. The loss of E-cadherin causes cells to get detached from the extracellular matrix and migrate. Loss of E-cadherin is a hallmark of almost all advanced stages of cancer resulting in metastasis (Roy and Datta 2019; Mendonsa et al. 2018).

Several driver genes identified by Stephens et al. (2012) code for transcription factors. These are proteins that initiate the transcription of a gene by attaching to the 5' UTR of the gene. The reduced expression of a transcription factor results in lower expressions for the transcription target genes, and overexpression has the opposite result, both of these are hallmarks in many cancers, as tumor suppressor genes tend to be under-expressed and oncogenes tend to be overexpressed. GATA3, MLL2, and MLL3 are important transcription factors. KRAS is a RAS family oncogene and codes for the K-Ras protein that participates in the RAS/MAPK extracellular signalling pathway [(Roy and Datta 2019), Chap. 2]. The RAS family genes are often mutated in several cancers, and Stephens et al. (2012) identified KRAS as a key driver gene for ER+ and ER− breast cancers.

We review the functions of some of the new driver genes identified by Stephens et al. (2012), e.g., AKT2, ARID1B, CASP8, CDN1B, MAP3K1, MAP3K13, NCOR, SMARCD1, and TBX3. AKT2 is an oncogene that codes for an important protein kinase responsible for phosphorylating several proteins. This gene is overexpressed in several cancers, including pancreatic, breast, and ovarian cancers. ARID1B is a chromatin remodelling gene. CASP8 codes for a protein that plays an important role in the apoptosis pathway. CDN1B codes for a protein that participates in cell cycle regulation. MAP3K1 codes for a protein kinase that is activated through the process of autophosphorylation. MAP3K13 is also a protein kinase gene. NCOR1 codes for a chromatin remodelling protein whose role is to condense the chromatin and deny access to transcription factors. SMARCD1 is also a chromatin remodelling gene, and TBX3 codes for a transcription factor.

Nik-Zainal et al. (2016) studied the genomic landscape of breast cancer by sequencing 560 breast tumors (556 females and 4 males) of ER+ and ER− patients. Many of their findings matched the findings of Stephens et al. (2012), and they also found other possible driver mutations in previously unknown genes. They found about 3.5 million somatic base substitutions, over 371,000 small indels (insertion and deletions of nucleotides) and over 77,000 rearrangements. They clustered the indels and somatic base substitutions in the exons of pre-mRNAs and found five new previously unknown cancer genes, FOXP1, XBP1, MLLT4, MED23, and ZFP36L1. However, the mutations indicated that these genes most probably act in breast cancer as recessive genes.

As we have seen in Chap. 1, breast cancer cells that have receptors for either estrogen (ER+), or progesterone (PR+), or both, are considered hormone receptor-positive (HR+). About 20% of ER+ patients do not respond to endocrine treatment, and another 30–40% develop endocrine resistance over the course of treatment. Endocrine resistance eventually results in metastasis and death (Nik-Zainal et al. 2016). Recently, Razavi et al. (2018) have studied the genomic landscape of endocrine resistant metastatic breast cancer. Their main findings are that an increasing number of genes involved in the MAPK pathway and estrogen receptor transcription get mutated with endocrine treatment. ERBB2 mutations and NF1 loss of function mutations were almost double in estrogen resistant tumors compared to estrogen non-resistant tumors. ERBB2 codes for a receptor tyrosine kinase of the epidermal growth factor (EGF) family, and NF1 codes for the neurofibromin protein that negatively regulates the RAS/MAPK pathway. They also found higher mutations compared to estrogen non-resistant tumors in MAPK pathway genes like KRAS and EGFR. The mutations were also much higher in estrogen receptor transcriptional regulator genes like TBX3, FOXA1, MYC, and CTCF.

2.4 Genomic Landscape of HER2-Positive Breast Cancer

The amplification and overexpression of the ERBB2 gene that codes for the HER2 protein (human epidermal growth factor receptor 2) have been observed in 15–30% of all invasive breast cancers. Most treatments target HER2-targeted agents and resistance to such treatments results in poor prognosis. Lesurf et al. (2017) studied the genomic landscape of HER2-positive breast cancer for understanding resistance to treatment. They performed RNA-seq of 42 tumor biopsies, and 50% of the patients achieved pCR (pathological complete response, when all signs of cancer have been removed after chemotherapy). The other 50% had residual disease (RD).

Lesurf et al. (2017) identified 196 pathogenic germline variants and 19 possible pathogenic germline variants by analyzing the normal genomes of the patients. The affected genes included BRCA2, MUTYH, ATM, FANCA, WNT10A, LRP5, PIK3R5, and TERT. MUTYH codes for a protein MYH glycosylase that is responsible for correcting errors during DNA replication. ATM codes for a protein that is responsible both for cell cycle regulation and DNA repair. FANCA codes for a protein that is responsible for cell cycle regulation in response to specific DNA damage. The Wnt signalling pathway activates gene expression regulation through extracellular signals [(Roy and Datta 2019), Chap. 2]. WNT10A codes for a key protein that participates in the Wnt signalling pathway. LRP5 codes for a co-receptor protein on the cell membrane that can recognize the extracellular signals in the Wnt pathway. PIK3R5 codes for a key protein in the PI3K/AKT/mTOR intracellular signalling pathway (Roy and Datta 2019, Chap. 2). Telomeres are repeated DNA sequences at the ends of chromosomes that protect the chromosomes from breaking down and fusing together. The lengths of telomeres shorten after each cell division cycle and a normal cell usually dies after a few cell division cycles. TERT encodes for the protein hTERT that is responsible for adding the telomere nucleotides at the end of chromosomes. Abnormal overexpression of TERT may make normal cells potentially immortal, as is the case for cancer cells.

Though previous studies (Kim et al. 2012; Norton et al. 2014) had reported significant germline polymorphisms in patients treated with chemotherapy and trastuzumab in the genes ABCB1, FCGR2A, FCGR2B, and FCGR3A. Lesurf et al. (2017) did not find any significant differences in these polymorphisms between pCR and RD patients. ABCB1 codes for a protein that transports various molecules across intra- and extracellular membranes. Overexpression of ABCB1 may be responsible for drug resistance of cancer cells, as the drug molecules are transported away and do not accumulate in the cancer cells (Wu et al. 2021).

Lesurf et al. (2017) also studied somatic mutations and found over 6600 somatic events, both single nucleotide variations (SNV) and small indels. The significant changes were observed only in PIK3CA, TP53, and MT-ND2 genes. MT-ND2 is a mitochondrial gene that codes for the NADH dehydrogenase 2 (ND2) protein. Mutation in this gene has been implicated in metastasis (Marco-Bruall et al. 2019). Lesurf et al. (2017) also investigated whether genetic pathways were enriched in SNV and indels. They found that several pathways were significantly enriched in SNV and indels for all 48 patients, including PI3K/AKT/mTOR, Wnt, p53, and TGFβ. TGFβ (Roy and Datta 2019, Chap. 2) is an extracellular signalling pathway that sends a signal to the nucleus for regulating gene expression through modulation of transcription factors.

2.5 Genomic Landscape of Basal-Like and Triple-Negative Breast Cancer

Basal-like breast cancer (BLBC) is an aggressive subtype of breast cancer with the characteristic that the epithelial cells at the basal or outer layer of the mammary gland express a robust cluster of genes (Troft and Cryns 2011). Triple-negative breast cancers (TNBC) are BLBC and negative for ER, PR, and HER2 (hence the name triple negative). Most basal-like tumors are TNBC, but approximately 25% of basal-like tumors are of other mRNA subtypes (The Cancer Genome Atlas Network 2012). The first attempt at determining the genomic landscape of triple-negative breast cancer was perhaps by Shah et al. (2012). They performed RNA-seq analysis of 104 primary TNBC tumors and identified PARK2, RB1, PTEN, and EGFR as the highest mutated genes. PARK2 codes for the gene parkin which is a ubiquitin ligase that has very important role in ubiquitination, the intracellular degradation of target molecules (Gupta et al. 2017). Loss of parkin due to mutation in PARK2 results in Parkinson's disease and loss of parkin may have important roles in carcinogenesis as a negative regulator of the PI3K/AKT/mTOR intracellular signalling pathway, responsible for cell cycle regulation. PARK2 depletion results in marked activation of this pathway and cell proliferation (Gupta et al. 2017). RB1 codes for the retinoblastoma protein or pRb, which is an important tumor suppressor protein. It is a negative regulator of the cell cycle, and cell cycle progresses when pRb is phosphorylated (Chinnam and Goodrich 2011). As a hall mark of cancer is cell proliferation, mutation in RB1 can be implicated for carcinogenesis. However, RB1 may have many other roles in carcinogenesis, including survival, differentiation, and senescence (Chinnam and Goodrich 2011).

The Cancer Genome Atlas Network (2012) studied the genomic landscape of both BLBC and TNBC along with other subtypes of breast cancer. They took samples from 76 TNBC, 81 basal-like and 65 both TNBC and basal-like tumor samples and used whole exome sequencing and other techniques. About 80% of BLBC showed a high frequency of TP53 mutations. It was also determined that another prominent basal-like feature is the loss of RB1 and BRCA1. PIK3CA was found to be the next most mutated gene. The top-five mutated genes in TNBC as reported by Shah et al. (2012) matched the top-five mutated genes for TNBC in this study (The Cancer Genome Atlas Network 2012) as well. In addition, high expression levels of FOXM1 and MYC, mutations in ATM, and inactivation of BRCA1 and BRCA2 were identified as features of basal-like tumors. FOXM1 is a proto-oncogene that codes for the FOXM1 protein. This protein plays very important roles in cell cycle progression. Upregulation of FOXM1 has been found in almost all solid human cancers, including breast, liver, prostrate, lung, brain, and colon (Kalathil et al. 2021). FOXM1 also may play important roles in invasion, migration, and angiogenesis (Kalathil et al. 2021).

In a recent study, Chen et al. (2020) performed an extensive study for identifying differentially expressed genes and key pathways in TNBC using a variety of bioinformatics analysis. They identified several previously unreported hub genes for TNBC, namely HSPA8, ESR1, ACTB, HSP90AA1, SRC, PPP2CA, and RPL4. HSPA8 codes for a heat shock protein that has many roles in cellular processes, including in apoptosis, autophagy, cell growth, and differentiation. It has been found to be significantly overexpressed in endometrial carcinoma (Shan et al. 2016), a type of cancer that begins in the uterus. ESR1 codes for an estrogen receptor and has been found to be frequently mutated in endocrine therapy resistant breast cancers (Zundelevich et al. 2020). ACTB codes for the protein beta-actin that is responsible for cell motility. ACTB is overexpressed in many cancers including renal, colorectal, liver, pancreatic, esophageal, melanoma, gastric, lung, prostate, ovarian, breast, leukemia, and lymphoma (Guo et al. 2013). Molecular chaperones are proteins that are involved in folding/unfolding of other proteins. HSP90AA1 is a molecular chaperone that is expressed widely in many cell types. HSP90AA1 may be responsible for stabilizing misfolded and overexpressed proteins, thus helping cancer cells to survive (Xiao et al. 2018). It has also been reported that HSP90AA1 is expressed in the extracellular environment, resulting in tumor progression and invasion of cancer cells (Xiao et al. 2018). A recent study has found that the expression levels of HSP90AA1 are a good predictor for onset and metastasis of breast cancer (Liu et al. 2021). SRC is a tyrosine kinase and phosphorylates specific tyrosine residues in several other tyrosine kinases. Overexpression of SRC has been implicated in the progression of many cancers (Irby and Yeatman 2000). Protein phosphatase 2A or PP2A is one of the main threonine/serine phosphatases in all eukaryotic cells and removes the phosphate group from many proteins (Janssens and Goris 2001). The suppression of PP2A has been implicated in carcinogenesis and tumorigenesis in several cancers (Junttila et al. 2007). PPP2CA codes for the alpha isoform of PP2A, and its suppression has been implicated in prostate cancer (Singh et al. 2008) and acute myeloid leukemia (Park et al. 2007). A recent study found that genetic variants in PPP2CA are associated with gastric cancer risk (Huang et al. 2017). The RPL4 gene codes for the protein 60S ribosomal protein L4 that is a constituent of the 60S subunit of the ribosome, the organelle where mRNA is translated into protein. It is known that the protein MDM2 has a crucial role in stabilizing the concentration of the p53 protein, one of the most important tumor suppressor proteins. The concentration of p53 increases in response to DNA damage, and overexpression of MDM2 reduces this concentration (Kubbutat et al. 1997). It has been reported recently that RPL4 directly binds to MDM2 and prevents MDM2-mediated ubiquitination of p53 (He et al. 2016).

2.6 Genomic Landscape of Ductal Carcinoma In Situ

Ductal carcinoma in situ (DCIS) represents about 80% of all in situ cancers and 20–25% of all breast cancers (van Seijen et al. 2019). The somatic copy number alterations (CNA) and single nucleotide variations (SNV) increased gradually as ductal carcinoma in situ progressed from hyperplasia to carcinoma in situ to invasive carcinoma (Afghahi et al. 2015). Afghahi et al. (2015) also found that many of these CNA and SNVs are common with previously reported ones for invasive breast carcinoma (IBC). They found 45% of the alterations were low frequency in the genes MLL3, PTEN, and GATA3; 27% were moderate frequency alterations in ERBB2 and CDH1; and about 27% were high frequency mutations in TP3 and PIK3CA. Lin et al. (2019) in a recent study also confirmed that many of these CNA and SNVs are common between DCIS and IBC. The most common mutations were in PIK3CA (34.4%), TP53 (18.4%), and GATA3. Mutations in PIK3CA kinase domain and progression of DCIS were inversely correlated. They concluded that there is no single genomic feature correlating DCIS to IBC. However, they found a gradual increase in ERBB2 amplification as IBC progressed (Lin et al. 2019).

Agahozo et al. (2019) studied the mutation levels and expression levels of PIK3CA and related genes in DCIS and IBC. As we have noted before, PIK3CA codes for the p110α protein which is a catalytic subunit of PI3K, a key member of the PI3K/AKT/mTOR intracellular signalling pathway. AKT is activated at the plasma membrane by phosphorylation of its amino acid residues (p-AKT) and can in turn phosphorylate many different proteins. PTEN is a tumor suppressor gene/protein and inhibits PI3K. Agahozo et al. (2019) detected a significantly higher variant allele frequency of PIK3CA in DCIS compared to adjacent IBC. The expression level of PTEN was much higher in DCIS compared to IBC when wild type PIK3CA expression was higher. However, PTEN expression was similar in DCIS and IBC when PIK3CA was predominantly mutated. The expression of p-AKT was similar in both IBC and DCIS.

Doebar et al. (2019) recently did a comparison of genomic landscape between DCIS and IBC. They analyzed whole exon sequences of four DCIS patients who were ER+ and HER2-negative and had IBC after surgical procedures. They identified 92 tumor specific variants of genes, and 52 of those overlapped between DCIS and IBC. All tumor specific variants present in IBC were also present in DCIS.

Petridis et al. (2016) used data from 38 studies, with over 37,000 controls, over 5000 cases of DCIS and over 24,000 cases of invasive ductal carcinoma (IDC). They confirmed that about 67% of the 76 previously known loci for predisposition to breast cancer were also associated with both DCIS and IBC. They also confirmed that most of the loci associated with ER-positive IDC were associated with ER-positive DCIS. They identified two SNPs near the CCND1 gene that were specific to low and medium grade DCIS, but not with IBC (Petridis et al. 2016). The CCND1 gene codes for the cyclin D1 protein that is important in cell cycle regulation. Mutations in CCND1 have been implicated in mantle cell lymphoma (Weistner et al. 2007) and endometrioid endometrial adenocarcinoma (Xu and Lin 2018).

2.7 Genomic Landscape of Invasive Lobular Carcinoma

Invasive lobular carcinoma (ILC) accounts for about 15% of all breast cancers and it is important to understand its genomic landscape for developing effective treatments. Ciriello et al. (2015) profiled over 800 breast tumors, including ILC, invasive ductal carcinoma (IDC), and both ILC and IDC. E-cadherin loss due to suppression of the CDH1 gene was identified by them as the major hallmark of ILC. They also identified mutations in PTEN, FOXA1, and TBX3 as essential features of ILC. Mixed ILC and IDC samples were either mostly ILC-like, or IDC-like, without any clear features for the mixed group. As mentioned before, FOXA1 and TBX3 are estrogen receptor transcriptional regulator genes.

Pereira et al. (2016) confirmed the findings in (Ciriello et al. 2015), in particular the loss of E-cadherin as a hallmark of ILC. McCart Reed et al. (2021) recently performed a meta-analysis of data from several prior studies on the genomic landscape of ILC, including TCGA (Berger et al. 2018), EURO (Desmedt et al. 2016), METABRIC (Pereira et al. 2016; Curtis et al. 2012), and RATHER (Michaut et al. 2016). These previous studies had different populations of grades 1/2/3 ILC patients and tumors. They found that the most common driver mutations for ILC are CDH1, FOXA1, ERBB2, ERBB3, PIK3CA, TP53, TBX3, PTEN, and FGFR2. It has been previously reported that mutations in CTNNA1 and AKT1 are important in ILC (de Groot et al. 2018; Teo et al. 2018), however McCart Reed et al. (2021) did not find significant levels of mutations in these two genes. ERBB3 codes for a receptor tyrosine kinase, a membrane bound protein that is a member of epidermal growth factor receptor family of proteins. ERBB3 is aberrantly overexpressed in many cancers, and it promotes tumor progression through interactions with other receptor tyrosine kinases (Ma et al. 2014). FGFR2 codes for the FGFR2 protein that is a fibroblast growth factor receptor and transmits signals from extracellular fibroblast growth factors for cell cycle regulation. CTNNA1 codes for the catenin alpha-1 protein that has important role in cell-to-cell adhesion. Loss of function of this protein has been implicated in metastatic gastric and breast cancers (Farengo et al. 2020).

2.8 Conclusion

We have reviewed the current understanding of the genomic landscapes of different types of breast cancers in this chapter. A summary of this review is presented in Table 2.1. These landscapes are often heterogeneous even between different cells within the same tumor and across tumors of the same type. Different genes may be mutated in different degrees in different cells. Still, some of the key driver genes have been identified for different types of breast cancers, including genes that are involved in signalling, cell cycle regulation, and cell adhesion. Some of the key pathways also have been identified in the studies reviewed in this chapter. Many of the genetic mutations identified in breast cancer have also been implicated in other cancers. We will discuss epigenetics of breast cancer in the next chapter, as aberrant epigenetic modulations play important roles in breast cancer along with mutations in genes.

Table 2.1 Summary of driver gene mutations in different breast cancer types

ER+, ER−	HER2-positive	Basal-like and triple negative	DCIS	IBC
BRCA1	BRCA2	PARK2, RB1	MLL3, PTEN	PTEN, CDH1
BRCA2	MUTYH	PTEN, EGFR	GATA3	FOXA1
PIK3CA, TP53	ATM, FANCA	BRCA1	ERBB2	ERBB2
AKT1, PTEN	WNT10A	PIK3CA	CDH1, TP3	ERBB3
RB1, GATA3	LRP5, PIK3R5	FOXM1, MYC	PIK3CA	PIK3CA, TP53
CDH1, KRAS	TERT	ATM, BRCA1	ATM, BRCA1	FGFR2, TBX3
MLL2	ABCB1	BRCA2	CCND1	CTNNA1
MAP2K4	FCGR2A	HSPA8, ESR1		AKT1
MLL3, ARID2	FCGR2B	ACTB		
ARID1A, APC	FCGR3A	HSP90AA1		
ASXL1		SRC, PPP2CA		
SETD2, NF1		RPL4		
SF3B1, STK11				
BAP1				
SMAD4				
NCOR1				
AKT2, CASP8				
ATID1B				
MAP3K1				
SMARCD1				
MAP3K13				
TBX3				
CDKN1B				

References

Afghahi A et al (2015) Chromosomal copy number alterations for associations of ductal carcinoma in situ with invasive breast cancer. Breast Cancer Res 17:108. https://doi.org/10.1186/s13058-015-0623-y

Agahozo MC et al (2019) PIK3CA mutations in ductal carcinoma in situ and adjacent invasive breast cancer. Endocr Relat Cancer 26:471–482

Anantha RW, Simhadri S, Foo TK, Miao S et al (2017) Functional and mutational landscapes of BRCA1 for homology-directed repair and therapy resistance. elife 6:e21350

Bau DT, Mau YC, Shen CY (2006) The role of BRCA1 in non-homologous end-joining. Cancer Lett 240(1):1–8. https://doi.org/10.1016/j.canlet.2005.08.003

Berger AC et al (2018) A comprehensive pan-cancer molecular study of gynecologic and breast cancers. Cancer Cell 33:690–705.e9

Chen CY, Chen J, He L, Stiles BL (2018) PTEN: tumor suppressor and metabolic regulator. Front Endocrinol 9:338. https://doi.org/10.3389/fendo.2018.00338

Chen J et al (2020) KEGG-expressed genes and pathways in triple negative breast cancer: protocol for a systematic review and data mining. Medicine (Baltimore) 99(18):e19986. https://doi.org/10.1097/MD.0000000000019986

Chial H (2008) Tumor suppressor (TS) genes and the two-hit hypothesis. Nature Education 1(1): 177

Chinnam M, Goodrich DW (2011) *RB1*, development, and cancer. Curr Top Dev Biol 94:129–169. https://doi.org/10.1016/B978-0-12-380916-2.00005-X

Ciriello G et al (2015) Comprehensive molecular portraits of invasive lobular breast cancer. Cell 163(2):506–519. https://doi.org/10.1016/j.cell.2015.09.033

Curtis C et al (2012) The genomic and transcriptomic architecture of 2,000 breast tumours reveals novel subgroups. Nature 486:346–352

de Groot JS et al (2018) Alphae-catenin is a candidate tumour suppressor for the development of E-cadherin-expressing lobular-type breast cancer. J Pathol 245:456–467

Densham RM, Morris JR (2017) The BRCA1 ubiquitin ligase function sets a new trend for remodelling in DNA repair. Nucleus 8:116–125

Desmedt C et al (2016) Genomic characterization of primary invasive lobular breast cancer. J Clin Oncol 34:1872–1881

Doebar SC et al (2019) Progression of ductal carcinoma in situ to invasive breast cancer: comparative genomic sequencing. Virchows Arch 474:247–251. https://doi.org/10.1007/s00428-018-2463-5

Farengo D et al (2020) Loss-of-function variants in CTNNA1 detected on multigene panel testing in individuals with gastric or breast cancer. Genet Med 22:840–846. https://doi.org/10.1038/s41436-020-0753-1

Guo C et al (2013) ACTB in cancer. Clin Chim Acta 417:39–44. https://doi.org/10.1016/j.cca.2012.12.012

Gupta A, Anjomani-Virmounni S, Koundourous N, Poulogiannis G (2017) *PARK2* loss promotes cancer progression via redox-mediated inactivation of PTEN. Mol Cell Oncol 4(6):e1329692. https://doi.org/10.1080/23723556.2017.1329692

He X, Li Y, Dai MS, Sun XX (2016) Ribosomal protein L4 is a novel regulator of the MDM2-p53 loop. Oncotarget 7(13):16217–16226. https://doi.org/10.18632/oncotarget.7479

Huang T et al (2017) Genetic variants in PPP2CA are associated with gastric cancer risk in a Chinese population. Sci Rep 7:11499. https://doi.org/10.1038/s41598-017-12040-z

Irby RB, Yeatman TJ (2000) Role of Src expression and activation in human cancer. Oncogene 19: 5636–5642. https://doi.org/10.1038/sj.onc.1203912

Janssens V, Goris J (2001) Protein phosphatase 2A: a highly regulated family of serine/threonine phosphatases implicated in cell growth and signalling. Biochem J 353:417–439. https://doi.org/10.1042/bj3530417

Junttila MR et al (2007) CIP2A inhibits PP2A in human malignancies. Cell 130:51–62. https://doi.org/10.1016/j.cell.2007.04.044

Kalathil D, John S, Nair AS (2021) FOXM1 and cancer: faulty cellular signaling derails homeostasis. Front Oncol 10. https://doi.org/10.3389/fonc.2020.626836

Kim JW et al (2012) ABCB1, FCGR2A, and FCGR3A poly-morphisms in patients with HER2-positive metastatic breast cancer who were treated with first-line taxane plus trastuzumab chemotherapy. Oncology 83:218–227

Knudson AG (2001) Two genetic hits (more or less) to cancer. Nat Rev Cancer 1:157–162

Kubbutat MH, Jones SN, Vousden KH (1997) Regulation of p53 stability by Mdm2. Nature 387: 299–303. https://doi.org/10.1038/387299a0

Lesurf R et al (2017) Genomic characterization of HER2-positive breast cancer and response to neoadjuvant trastuzumab and chemotherapy—results from the ACOSOG Z1041 (Alliance) trial. Ann Oncol 28:1070–1077. https://doi.org/10.1093/annonc/mdx048

Lin CY et al (2019) Genomic landscape of ductal carcinoma in situ and association with progression. Breast Cancer Res Treat 178(2):307–316. https://doi.org/10.1007/s10549-019-05401-x

Liu H et al (2021) Plasma HSP90AA1 predicts the risk of breast cancer onset and distant metastasis. Front Cell Dev Biol 9:639596. https://doi.org/10.3389/fcell.2021.639596

Ma J, Lyu H, Huang J, Liu B (2014) Targeting of erbB3 receptor to overcome resistance in cancer treatment. Mol Cancer 13:105. https://doi.org/10.1186/1476-4598-13-105

Marco-Bruall J et al (2019) Mutations in the *ND2* subunit of mitochondrial complex I are sufficient to confer increased tumorigenic and metastatic potential to cancer cells. Cancers (Basel) 11(7): 1027. https://doi.org/10.3390/cancers11071027

McCart Reed AE et al (2021) The genomic landscape of lobular breast cancer. Cancers (Basel) 13: 1950. https://doi.org/10.3390/cancers13081950

Mendonsa AM, Na TY, Gambiner BM (2018) E-cadherin in contact inhibition and cancer. Oncogene 37:4769–4780

Meric-Bernstam F (2007) Heterogenic loss of *BRCA* in breast cancer: the "two-hit" hypothesis takes a hit. Ann Surg Oncol 14:2428–2429

Michaut M et al (2016) Integration of genomic, transcriptomic and proteomic data identifies two biologically distinct subtypes of invasive lobular breast cancer. Sci Rep 6:18517

Mullan PB, Quinn JE, Harkin DP (2006) The role of BRCA1 in transcriptional regulation and cell cycle control. Oncogene 25:5854–5863

NCBI (n.d.) NCBI gene resources. https://www.ncbi.nlm.nih.gov/gene

Nik-Zainal S et al (2016) Landscape of somatic mutations in 560 breast cancer whole-genome sequences. Nature 534:47–54. https://doi.org/10.1038/nature17676

Norton N et al (2014) Association studies of Fc gamma receptor polymorphisms with outcome in HER2þ breast cancer patients treated with trastuzumab in NCCTG (Alliance) trial N9831. Cancer Immunol Res 2:962–969

Ohta T, Sato K, Wu W (2011) The BRCA1 ubiquitin ligase and homologous recombination repair. FEBS Lett 585:2836–2844

Oliver AW, Swift S, Lord CJ, Ashworth A, Pearl LH (2009) Structural basis for recruitment of BRCA2 by PALB2. EMBO Rep 10:990–996

Park MH et al (2007) Gene expression profile related to prognosis of acute myeloid leukemia. Oncol Rep 18:1395–1402

Pereira B et al (2016) The somatic mutation profiles of 2, 433 breast cancers refine their genomic and transcriptomic landscapes. Nat Commun 7:11479. https://doi.org/10.1038/ncomms11479

Petridis C et al (2016) Genetic predisposition to ductal carcinoma in situ of the breast. Breast Cancer Res 18:22. https://doi.org/10.1186/s13058-016-0675-7

Phillips T, Shaw K (2008) Chromatin remodeling in eukaryotes. Nature Education 1(1):209

Razavi P et al (2018) The genomic landscape of endocrine-resistant advanced breast cancers. Cancer Cell 34(3):427–438. https://doi.org/10.1016/j.ccell.2018.08.008

Roy M, Datta A (2019) Cancer genetics and therapeutics: focus on phytochemicals. Springer, Cham

Roy R, Chun J, Powell SN (2011) BRCA1 and BRCA2: different roles in a common pathway of genome protection. Nat Rev Cancer 12:68–78

Savage KI, Harkin DP (2015) BRCA1, a 'complex' protein involved in the maintenance of genomic stability. FEBS J 282:630–646

Semmler L, Reiter-Brennan C, Klein A (2019) BRCA1 and breast cancer: a review of the underlying mechanisms resulting in the tissue-specific tumorigenesis in mutation carriers. J Breast Cancer 22(1):1–14. https://doi.org/10.4048/jbc.2019.22.e6

Shah SP et al (2012) The clonal and mutational evolution spectrum of primary triple-negative breast cancers. Nature 486:395–399

Shan N, Zhou W, Zhang S, Zhang Y (2016) Identification of HSPA8 as a candidate biomarker for endometrial carcinoma by using iTRAQ-based proteomic analysis. Onco Targets and Ther 9: 2169–2179. https://doi.org/10.2147/OTT.S97983

Singh AP et al (2008) Genome-wide expression profiling reveals transcriptomic variation and perturbed gene networks in androgen-dependent and androgen-independent prostate cancer cells. Cancer Lett 259:28–38. https://doi.org/10.1016/j.canlet.2007.09.018

Spatz MA, US National Library of Medicine (2004) Genetics home reference. J Med Libr Assoc 92(2):282–283. https://ghr.nlm.nih.gov/gene

Stephens PJ et al (2012) The landscape of cancer genes and mutational processes in breast cancer. Nature 486(7403):400–404. https://doi.org/10.1038/nature11017

Teo K et al (2018) E-cadherin loss induces targetable autocrine activation of growth factor signalling in lobular breast cancer. Sci Rep 8:15454

The Cancer Genome Atlas Network (2012) Comprehensive molecular portraits of human breast tumours. Nature 490:61–70. https://doi.org/10.1038/nature11412

Troft DJ, Cryns VL (2011) Minireview: basal-like breast cancer: from molecular profiles to targeted therapies. Mol Endocrinol 25(2):199–211. https://doi.org/10.1210/me.2010-0164

van Seijen M et al (2019) Ductal carcinoma in situ: to treat or not to treat, that is the question. Br J Cancer 121:285–292. https://doi.org/10.1038/s41416-019-0478-6

Weistner A et al (2007) Point mutations and genomic deletions in *CCND1* create stable truncated cyclin D1 mRNAs that are associated with increased proliferation rate and shorter survival. Blood 109(11):4599–4606. https://doi.org/10.1182/blood-2006-08-039859

Wu ZX et al (2021) Elevated ABCB1 expression confers acquired resistance to Aurora kinase inhibitor GSK-1070916 in cancer cells. Front Pharmacol 11:615824. https://doi.org/10.3389/fphar.2020.615824

Xiao X et al (2018) HSP90AA1-mediated autophagy promotes drug resistance in osteosarcoma. J Exp Clin Cancer Res 37(201):201. https://doi.org/10.1186/s13046-018-0880-6

Xu J, Lin DI (2018) Oncogenic c-terminal cyclin D1 (CCND1) mutations are enriched in endometrioid endometrial adenocarcinomas. PLoS One 13(7):e0199688. https://doi.org/10.1371/journal.pone.0199688

Zhang F, Ma J, Wu J, Ye L et al (2009) PALB2 links BRCA1 and BRCA2 in the DNA-damage response. Curr Biol 19(6):524–529

Zundelevich A et al (2020) *ESR1* mutations are frequent in newly diagnosed metastatic and loco-regional recurrence of endocrine-treated breast cancer and carry worse prognosis. Breast Cancer Res 22:16. https://doi.org/10.1186/s13058-020-1246-5

3 Epigenetics of Breast Cancer

3.1 Introduction

Cancer is primarily a genetic disease as mutated genes result in malfunctional proteins, disrupting the normal functioning of cells. We have seen in Chap. 2 that scores of genes are mutated in different types of breast cancer. It is not possible to reverse these mutations and the only option is to destroy heavily mutated cells through radiotherapy, preventing the proliferation of heavily mutated cells. The central dogma of biology is that genes are transcribed to pre-mRNAs, introns from pre-mRNAs are spliced out to form mRNAs, and mRNAs are translated to proteins. Epigenetics is a higher level of control that can control transcription to modulate the concentrations of proteins, control the translation of mRNAs to proteins, and also modulate the functions of proteins. The epigenetic agents are also proteins that have structural conformations or binding domains that they use to bind to other proteins and molecules for performing their functions. These binding domains are often targets of drugs for inhibiting their functions, a drug molecule if bound to a functional domain of an epigenetic agent may prevent the protein from performing its normal functions.

We will first discuss the epigenetic modulations like histone methylation/demethylation, histone acetylation/deacetylation, and miRNA silencing. We will then discuss the specific epigenetic modulations that have been implicated in initiation, progression, and metastasis of different types of breast cancers.

3.2 A Review of Epigenetic Modulations

We first discuss transcriptional epigenetic control, namely histone modification and DNA methylation. Our discussion is based mainly on the book by Alberts et al. (2015). We then discuss post-transcriptional control of miRNA silencing.

M. Roy et al., *Genetics and Epigenetics of Breast Cancer*,
https://doi.org/10.1007/978-981-19-9925-3_3

3.2.1 Histone Modification

Except for mitochondrial DNA, most of the DNA in eukaryotic cells are tightly packed inside the nucleus. The DNA is wrapped around histone proteins resulting in a complex called chromatin. Histones are small positively charged proteins H1, H2A, H2B, H3, and H4, and the phosphate group of the phosphate-sugar backbone of DNA is negatively charged. Hence, DNA is packed tightly around histones (Annunziato 2008). Eight histone proteins constitute a nucleosome core particle, and chromatins are made of nucleosome core particles. Linker DNA connects two consecutive nucleosome core particles (Annunziato 2008). For a better understanding, this has been pictorially explained in Fig. 3.1.

Histone acetylation is the mechanism through which DNA is unpacked to give access to transcription factors to the transcription start regions of genes, for transcription to pre-mRNA. Histone deacetylation is the reverse process through which the DNA is packed tighter making a gene inaccessible by transcription factors. A histone protein has a core fold and an N-terminal tail. Processes like acetylation, methylation, and phosphorylation destabilize the histone structure by modifying these tails and allow access to transcription factors. Histone N-terminal ends have COOH and NH2 tails, and the NH2 tail is affected due to histone acetylation. Histone acetyltransferases (HAT) are proteins responsible for adding acetyl groups to the NH2 tails, and acetylation neutralizes the positive charge of the lysine amino acid residue by adding a negative charge. As a result, the histone cannot neutralize the negative charge of the DNA and the DNA is loosened. HATs activate transcription, and histone deacetylase (HDAC) enzymes can remove the acetyl groups to restore the positive charge of the lysine amino acid residue, packing DNA tightly again.

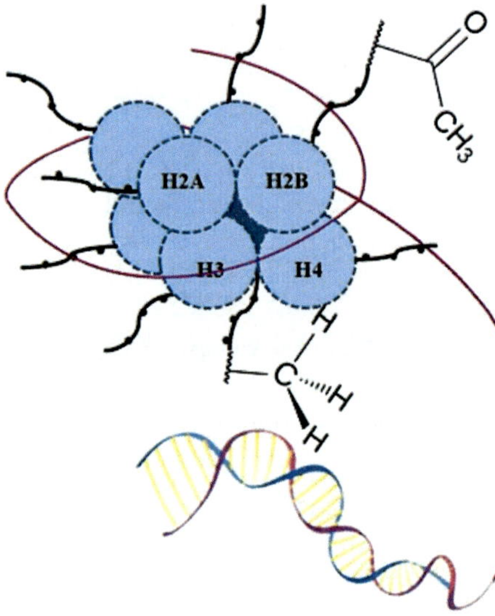

Fig. 3.1 Histone modifications like methylation, acetylation, and phosphorylation affect the N-terminal tail, loosening the DNA

Histone methylation and demethylation are finer epigenetic controls compared to histone acetylation and deacetylation. Histone methylation can increase or decrease the level of transcription of genes depending on how many amino acids are methylated. Histone methyltransferases (HMT) are the enzymes that attach methyl groups to lysine amino acid residues. Unlike histone acetylation, lysine residues both in the tail and histone core may be methylated, and the methyl group does not alter the charge. Moreover, a single lysine residue may be mono, di, or tri methylated, e.g., the fourth lysine residue on H3 (H3K4) has three methylation states, H3K4me, H3K4me2, and H3K4me3, indicating mono-, di-, and tri-methylation. Proteins with PHD (plant homeodomain) finger domains can accurately count the number of methylations and recruit chromatin remodelling proteins that loosen the DNA to give access to transcription factors (Musselman and Kutateladze 2011). Histone demethylases (HDM) can remove the methyl groups from amino acid residues to tighten the wrapping of DNA.

Histone phosphatases are the enzymes that attach a phosphate group to amino acid residues at the histone tail, neutralizing the negative charge. As a result, the negative charge on the DNA is not neutralized, and DNA is loosened giving access to transcription factors. Histone phosphorylation plays an important role in DNA damage response, and phosphorylated H2A demarcates the chromatin domains around DNA breakage sites (Rossetto et al. 2012).

3.2.2 DNA Methylation

DNA methylation is an important transcriptional epigenetic control. A family of enzymes called DNA methyltransferases add methyl groups mainly to adenine and cytosine, at the 5′ promoter regions of transcription start sites of genes, thus preventing transcription factors access to the transcription start sites. Figure 3.2 clearly demonstrates the DNA methylation process. Many vertebrate genes have CpG islands in the promoter regions of transcription start sites. These are repeated sequences of cytosine and guanine from 5′ to 3′ end of DNA, each pair connected by a phosphate group (Deaton and Bird 2011), as any two consecutive bases on the DNA are connected by a phosphate group. The CpG rich parts of the human genome are much lower compared to what it should be due to random chance (International Human Genome Sequencing Consortium 2001). Approximately 70% of the genes in the human genome have CpG islands in the promoter regions near transcriptional start sites (Deaton and Bird 2011). CpG islands are usually hypomethylated in normal cells, though the overall genome is hypermethylated. However, the CpG islands associated with the promoter regions near the transcription start sites of tumor suppressor genes are hypermethylated in many cancers. Several cancer drugs target DNA methyltransferases for preventing this hypermethylation (International Human Genome Sequencing Consortium 2001).

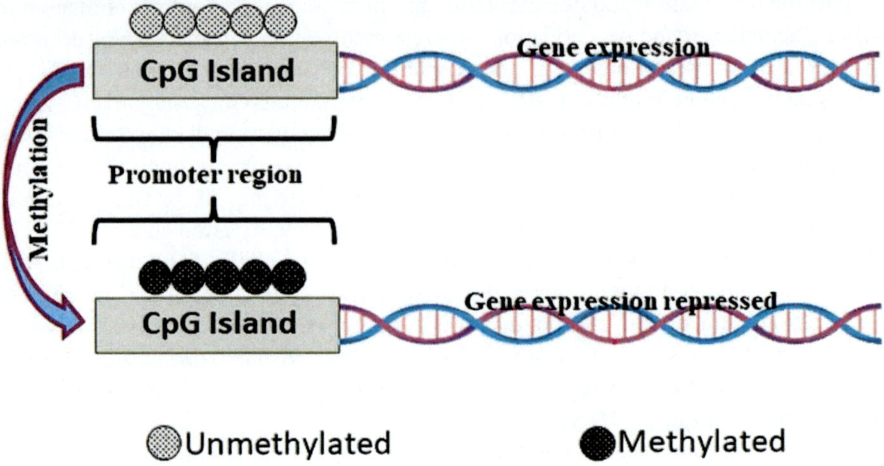

Fig. 3.2 DNA methylation of the CpG islands of the promoter regions of genes prevents transcription factors from binding at the promoter regions and prevents transcription

3.2.3 miRNA Silencing

miRNA silencing is a post-transcriptional silencing of mRNAs, which is explained in Fig. 3.3. miRNAs are small noncoding RNAs that are typically 18–25 nt long, they are transcribed from specific genes, but they are not translated to proteins. They bind to the 3′ noncoding regions of mRNA (3′ UTR) and prevent the translation of mRNAs to proteins. miRNA silencing is an ancient epigenetic control mechanism as many miRNA families are highly conserved in eukaryotic cells.

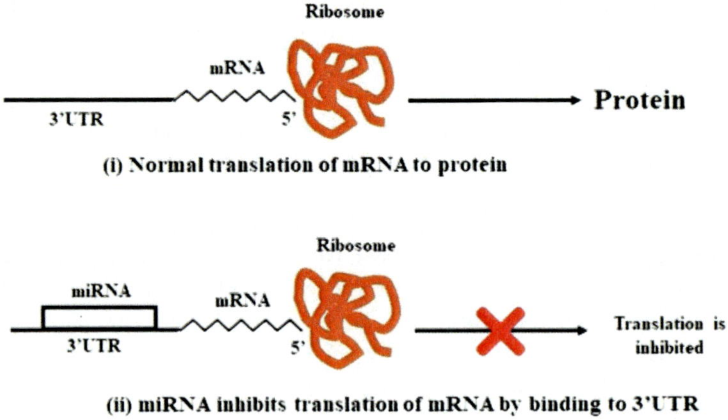

Fig. 3.3 Post-translational silencing by miRNA. (**a**) The mRNA is translated to a protein in normal circumstances; (**b**) binding of miRNA at the 3′ UTR of mRNA inhibits the translation of mRNA to protein

There is emerging evidence that DNA methylation, histone acetylation, and miRNA expression maintain a fine balance in normal cells, and this balance is disrupted in cancer cells including in breast cancer cells. We will discuss this later in this chapter. Approximately half of all known human miRNA genes are targeted for CpG methylation, and histone acetylation is the main mechanism for controlling the transcription of many miRNA genes. miRNAs on the other hand modulate the translation of histone deacetylases and DNA methyl transferases. Hence there is a significant level of cross regulations between different epigenetic control mechanisms. Different miRNAs act both as oncogenes and tumor suppressor genes. Some miRNAs play important roles in epithelial–mesenchymal transition (EMT) that causes cancer metastasis.

3.2.4 Fundamental Aspects of Epigenetics of Cancer

We first discuss some fundamental aspects of epigenetic control that has been observed in many cancers, before discussing epigenetics of different types of breast cancers. Our discussion is based on the review paper by Jones and Baylin (2007).

The overall genome is hypermethylated but the promoter regions near the transcription start sites containing CpG islands are hypomethylated in normal cells. However, the situation is just the opposite in many neoplastic cells. The CpG islands are often hypermethylated, and the genome is overall hypomethylated. The hypermethylation of CpG islands occurs in clusters of genes, some of those may be tumor suppressor genes, and as a result, a neoplastic cell gets a distinct growth advantage due to the suppression of tumor suppressor genes. CpG dinucleotides are also present in coding regions of genes, and cytosines are prone to mutation by external agents like ultraviolet (UV) rays or environmental agents when methylated, cytosine is mutated to thymine. These point mutations are the reason that CpG dinucleotides are underrepresented in the human genome (Jones and Baylin 2007). The TP53 gene is one of the most important tumor suppressor genes and highly mutated in most neoplastic cells, including those in all kinds of breast cancer. Many of the point mutations in TP53 are due to mutation of C to T in CpG dinucleotides in the coding part of TP53 (Baugh et al. 2017).

We have seen in Chap. 2 that Knudson's two-hit hypothesis is not applicable for the tumor suppressor gene BRCA1. Carcinogenesis may occur even if one copy of BRCA1 is mutated. However, the other copy is silenced by CpG island hypermethylation in the promoter region near the transcription start site. Furthermore, this hypermethylation does not occur for the mutated copy of BRCA1 (Jones and Baylin 2007).

Other chromatin modification processes are also responsible for gene expression regulation. Covalent histone modifications are also very important for gene regulation (Jones and Baylin 2007). Histone deacetylation and methylation act together in these processes. For example, histone acetylation and methylation of H3K9 and H3K27 participate in gene silencing (Jenuwein 2006). The interplay between histone covalent modification, histone methylation, and nucleosome remodelling—all are important factors for heritable gene silencing. The key enzymes in these processes are DNA methyltransferases, histone methyltransferases, histone deacetylases, and nucleosome remodelling factors (Jones and Baylin 2007).

Specific genome wide acetylation and methylation patterns also have been observed in many tumor cells. For example, acetylation is lost at H4K16 (the 16th lysine residue of histone H4) and tri-methylation of H4K20 (Fraga et al. 2005). The expression level of PRC2 (polycomb repressor complex 2) is high in several human cancers. This complex contains methyltransferases that methylate H3K9 and H3K27.

Epigenetic abnormalities may play important roles in cancer initiation and early stages of carcinogenesis. In particular, the hypermethylation of key tumor suppressor genes may affect several pathways that have been implicated in carcinogenesis. Studies on early stages of colon cancer have identified hypermethylation of several genes that affect important pathways, e.g., SFRP1, SFRP2, SFRP4 and SFRP5 affect the Wnt signalling pathway, MLH1 affects DNA mismatch repair pathway, CDKN2A affects cell cycle regulation, GATA4, GATA5, TFF1, TFF2, TFF3, and INHA affect epithelial cell differentiation, HIC1 affects p53-mediated DNA damage response, and TIMP3 affects cell invasion (Baylin and Ohm 2006).

3.3 Epigenetics of ER+ and ER− Breast Cancers

Hypermethylation of the promoter regions near transcription start sites and silencing of genes is a common occurrence in breast cancer (Pathiraja et al. 2010). These genes include BRCA1, a DNA repair gene; TMS1 and HOXA5, proapoptotic genes, HOXA5 also inhibits angiogenesis (Arderiu et al. 2007); CDH1 and CDH13, genes involved in cell adhesion and prevention of metastasis; RASSF1A and p16, genes involved in cell cycle inhibition; and GSTP1, a gene that encodes a metabolic enzyme (Pathiraja et al. 2010). High throughput and genome wide studies have identified several other hypermethylated genes including APC, BIN1, BMP6, CST6, ESRb, P21, and TIMP3 (Radpour et al. 2009). EZH2 and SUZ12 are overexpressed in many cancers, including breast cancers in early stages, and mark other genes, e.g., HoxA9 for methylation (Reynolds et al. 2006).

Global histone modifications have been found to be correlated with tumor progression and prognosis in breast cancer. Genome wide higher levels of methylation and acetylation usually correlate with better prognosis and outcomes for patients, mostly in luminal like tumors. Medium to lower levels of methylation and acetylation result in less favorable outcomes in basal-like, triple negative and HER2-positive tumors (Elsheikh et al. 2009). The histone acetyltransferase hMOF is downregulated in primary breast cancers and medulloblastoma (Pfister et al. 2008), a direct cause for low levels of acetylation.

Estrogen receptor 1 (ESR1) gene has low expression levels in ER- breast cancer cell lines. It has been reported that the CpG islands near the 5′ promoter region near transcription start site of ESR1 are hypermethylated in ER- cell lines (Ottaviano et al. 1994). MCF7 is a breast cancer cell line that is ER+ and estrogen-ablation resistant.

The CpG islands in the promoter region of this cell line are hypomethylated (Sogon et al. 2007). The interplay between histone modification, DNA methylation, and chromatin remodelling has been reported for ESR1 silencing (Rountree et al. 2000). Mutations, indels, and polymorphisms are rare in ESR1 gene, and the above results point to mainly epigenetic silencing for its low expression levels in ER− breast cancer.

Gao et al. (2015) did an integrative study of epigenomic and transcriptomic (using RNA-Seq) study of ER+ breast cancer. They investigated the variation of epigenomic silencing between the two subtypes of ER+ breast cancer, luminal-A and luminal-B. They concluded that these two subtypes are epigenetically similar in terms of DNA methylation, but the luminal-B subtype exhibits higher variations in DNA methylation. Gao et al. (2015) also concluded that DNA methylation of the genome in ER+ breast cancer is coordinated, different genes within the same pathways are methylated in coordination. Disruption of the Wnt signalling pathway results in initiation and progression of breast cancer (Veeck et al. 2008). The expression levels of Wnt pathway inhibitors like SFRP1, WIF1, WNT1, FZD9, and FZD8 are downregulated in both luminal-A and luminal-B subtypes due to hypermethylation of the 5′ promoter regions (Gao et al. 2015). FST and FSTL1 are two important inhibitors of the TGF-β signalling pathway and control cellular proliferation (Chan et al. 2009). The promoter regions of these two genes are also hypermethylated in ER+ breast cancer resulting in their under-expression, while TGFB1 and TGFB3 were overexpressed significantly (Gao et al. 2015). The protein BMP6 is a potential breast tumor suppressor (Hu et al. 2013), and its promoter region was also hypermethylated (Gao et al. 2015). The EDN2 gene codes for the protein Endothelin-2 that has been implicated in the progression of breast cancer (Ling et al. 2013). Gao et al. (2015) found it is hypomethylated and overexpressed in ER+ breast cancer. Underexpression of EDN3 has been implicated in breast cancer (Wiesmann et al. 2009), and EDN3 5′ promoter region was found to be hypermethylated (Gao et al. 2015).

3.4 Epigenetic Landscape of HER2-Positive Breast Cancer

In an early epigenetic study, Terada et al. (2009) identified an association between HER2-positive breast cancer progression and a higher number of genes silenced through CpG island methylation. In particular, they identified silencing of BRCA1, CDKN2A, and CDH1.

Lindqvist et al. (2014) performed a whole genome methylation analysis of 17 HER2-positive tumors and compared with normal samples. DNA methylation was altered in several groups of genes, including in genes responsible for differentiation, transcription, and multicellular development. They found over 144,000 CpGs are differentially methylated in HER2-positive tumor cells compared to normal cells. 1294 CpGs were significantly hypomethylated, 5519 CpGs were significantly hypermethylated, and hypermethylation was mostly concentrated in CpG islands in the promoter regions. Homeobox family of genes are transcription factors that regulate the transcription of other genes during early development. They identified several genes that were not associated with HER2-positive breast cancer earlier, DBX1, NKX2–6, and SIX6 are the new homeobox family of genes that were hypermethylated in HER2-positive breast cancer. In all, the CpG islands of promoter regions of 22 homeobox genes were hypermethylated. Hypomethylation of six homeobox genes were in the coding regions. The analysis of important genes affecting pathways revealed that the Wnt and PI3K/AKT pathways were most affected due to methylation (Lindqvist et al. 2014). Lindqvist et al. (2012) studied the methylation pattern of the 5′ promoter region of the SLC25A43 gene that undergoes loss of heterozygosity (copy from one parent is deleted) in HER2-positive breast cancer, though it is not mutated. They studied the methylation status with and without loss of heterozygosity for HER2-positive and HER2-negative tumor cells. Cancer tissues with no deletion of the SLC25A43 gene had higher methylation of the CpG islands at the 5′ promoter region, whereas the methylation was lower when one copy was deleted (Lindqvist et al. 2012).

Yamaguchi et al. (Singla et al. 2017; Yamaguchi et al. 2015) studied 24 HER2-positive breast cancer tissues for both genetic and epigenetic markers. They found that the negative regulator genes of the Wnt pathway, DKK3 and SFRP1 were hypermethylated in nine breast cancer cells, and IGFBP7, a p53 pathway regulatory gene was hypermethylated in all HER2-positive samples.

Misra et al. (2001) reported higher levels of acetylation of H3 and H4 and higher levels of methylation of H3 in HER2-positive cells. Also, the histone deacetylases were downregulated in these cells. However, there are also reports that histone methylation can repress HER2 expression levels in HER2-positive cells, by H3K9me2 (dimethylation of the ninth lysine residue in H3) (Falahi et al. 2013). Gene amplification is a process when cancer cells produce multiple copies of a gene due to aberrant signalling, and HER2-positive cells undergo amplification of the HER2 (also known as ERBB2 or ErbB2) gene, resulting in overexpression of the HER2 protein. Mungamuri et al. (2013) studied histone modification in HER2-positive cells in the absence of gene amplification and reported that H3K4me3 mark in the promoter region of HER2 is necessary for overexpression of HER2, and further the H3K9ac mark is necessary for receptor amplified tumors (Mungamuri et al. 2013). Li et al. (2014) confirmed the H3K4me3 mark in HER2-positive breast carcinoma and also reported the H3K27ac mark.

Persson et al. (2011) studied the genome wide miRNA expression profile for HER2-positive breast cancer. They identified hundreds of miRNAs expressed in HER2-positive cells, adjacent cells to HER2-positive cells, and normal breast cells of HER2-positive patients. The sets of miRNAs expressed in these three cell lines were largely overlapping, however some of the miRNAs were overexpressed in HER2-positive cells. In particular, Persson et al. (2011) identified two miRNA genes in the intron of the HER2 gene, miR-4728-5p, and miR-4728-3p, and miR-4728-3p was overexpressed along with the overexpression of HER2. They conjectured that miR-4728-3p may be involved in a feedback loop for overexpression of HER2, as the predicted targets of miR-4728-3p are downstream targets of HER2 (Grimson et al. 2007).

Castilla et al. (2012) studied the modulation of miRNA-200 family in several distinct breast cancer phenotypes, including ER+, HER2-positive, and triple negative and metaplastic breast carcinoma (that has both epithelial and mesenchymal cells). miR-200f had significantly reduced expression in metaplastic breast carcinoma. HER2-positive and triple-negative samples also had significantly lower expression levels of miR-200f compared to ER+ samples. It has been reported that miR-200f has important roles in controlling Epithelial to Mesenchymal Transition (EMT) (Sossey-Alaoui et al. 2009). Wee et al. (2012) studied miR-200b expression in breast cancer. They found a novel promoter region of miR-200b precursor, Hsa-miR-200b, rich in CpG dinucleotides and hypomethylation of this promoter region resulted in gain of HER2 and androgen expression. They suggested that potential methylation of this promoter could be an approach for controlling HER2-positive breast cancer. miR-21 expression level was high in advanced and large HER2-positive tumors (Lee et al. 2011). Higher expression levels of miR-210 have been reported in metastatic HER2-postive breast cancer (Jung et al. 2012). Overexpression of miR-155 has been observed in advanced HER2-positive breast cancer (Nassar et al. 2014). Overexpression of miR10-b and miR-19a was found in the serum of metastatic HER2-positive patients (Anfossi et al. 2014). Two miRNAs, miR-125b and let-7c have been found to directly target HER2 (Bailey et al. 2015).

3.5 Epigenetic Landscape of Basal-Like and Triple Negative Breast Cancer

Stirzaker et al. (2015) did a comprehensive genome wide study of methylation patterns in triple negative breast cancer (TNBC) tissues in comparison to normal breast tissues. They identified three different TNBC subtypes based on methylation clusters correlated with better or worse prognosis. This study identified hypermethylation events mostly in the CpG islands of promoter regions of genes. Twelve hypermethylated genes were both downregulated and mutated, including SEMA5A and ROBO3 (Stirzaker et al. 2015; Temian et al. 2018), genes involved in the axon guidance pathway during development. Aberration of this pathway has been implicated in breast cancer initiation, progression, and angiogenesis (Harburg and Hinck 2011). Hypermethylation of the promoter regions of seven genes in this pathway was identified by Stirzaker et al. (2015). Patients with hypomethylated profiles had better survival rate after diagnosis compared to heavily hypermethylated subtype, whereas the medium hypermethylated group had the worst survival rate.

Branham et al. (2012) identified hypermethylated promoter regions in 69 cancer-related genes. They identified TNBC specific hypermethylation of five genes, MGMT, RB, CDKN2B, CD44 and p73, and non-methylation of 11 genes (Branham et al. 2012). MMR genes, MGMT and ID4 had the strongest association with TNBC. MMR genes participate in the mismatch repair pathway for correcting insertion, deletion errors during DNA replication. MGMT also encodes a DNA repair protein. Branham et al. (2012) did not find any significant differences in methylation profiles of BRCA1 and BRCA2 genes in TNBC, compared to non-TNBC tissues. ID4 is a negative regulator of BRCA1, and it was one of the non-methylated genes. Hence an elevated expression level of ID4 may negatively regulate BRCA1, a tumor suppressor gene. It has been observed that the promoter region of BRCA1 in TNBC is hypermethylated if BRCA1 is not mutated (Lips et al. 2013).

Mathe et al. (2016) studied the correlation between DNA methylation and progression of TNBC. They compared the methylation of primary tumor and lymph node metastasis in a whole genome study. Out of 38 TNBC specific genes, 42% had significantly altered methylation, either hypomethylation or hypermethylation, compared to normal cells. They also identified aberrant methylation of 18 TNBC specific genes in lymph node metastasis. Further, Mathe et al. (2016) did a mapping of the aberrantly methylated genes to pathways. The hypomethylated genes were associated with several pathways including MAPK, axon guidance, and Rap 1 signalling pathways. The hypermethylated genes were associated with several pathways including PI3K/AKT and signalling pathways regulating pluripotency of stem cells (Mathe et al. 2016). Aberrant methylation of genes associated with stem cell properties have been reported by Kagara et al. (2012), including hypomethylation of CD44, CD133, and MSI1.

Holm et al. (2010) studied the differences of methylation patterns between basal-like breast cancer and luminal-A and luminal-B subtypes. They observed significantly different methylation between subtypes, basal-like tumors were least frequently methylated and luminal B tumors most frequently methylated. Luminal B subtype had more methylated targets of the polycomb repressor complex and embryonic stem cells (Holm et al. 2010). However, there are reports of a hypermethylated subgroup of basal-like breast cancer (Stirzaker et al. 2015; Hughes et al. 2013), where a set of genes are concurrently hypermethylated and silenced. These genes include CST6, CDH1, LCN2, CEACAM6, ESR1, and SCNN1A and are hypermethylated mainly due to the overexpression of DNMT3b, a DNA methyltransferase (Roll et al. 2008).

Xi et al. (2018) compared histone modification profiles between different subtypes of breast cancer by studying eight key histone marks, namely H3K79me2, H3K36me3, H3K27ac, H3K27me3, H3K9ac, H3K9me3, H3K4me3, and H3K4me1 across several cell lines including 4 TNBC cell lines. The TNBC cell lines had distinct H3K36me3 patterns, and AFAP1-AS1 (a long noncoding RNA) was marked by H3K4me3 and H3K79me2 modifications. Elevated expression level of AFAP1-AS1 has been observed earlier in different cancers (Dianatpur et al. 2018). AFAP1-AS1 may have roles in EMT as well, as the depletion of AFAP1-AS1 by small interfering RNA resulted in decrease in proliferation in TNBC cell lines (Xi et al. 2018). Gregoire et al. (2016) have reported the role of histone modification on EMT in basal-like cell line MDA-MB-231. E-cadherin expression increased by downregulating H3K79 methylator DOT1L, histone acetyltransferase KAT5, and histone methyltransferase G9a, and an epithelial phenotype with low invasive capacity was also promoted (Gregoire et al. 2016). Inhibition of bromodomain and extra-terminal family of proteins (BET) has antitumoral effects in solid tumors including TNBC (Shu et al. 2016). This family of proteins recognize acetylated lysine residues in histones.

In another study of histone modification on basal-like cancer, medium to low levels of lysine acetylation at H3K9ac, H3K18ac, H4K12ac, lysine methylation at H3K4me2, H4K20me3, and arginine methylation at H4R3me2 were reported for basal-like carcinomas with poor prognosis (Elsheikh et al. 2009).

MicroRNAs miR-10b and miR-26a have been identified as potential biomarkers for TNBC (Fkih M'hamed et al. 2015), downregulating the expression of BRCA1. miR-146a is overexpressed in TNBC cells without affecting the expression level of BRCA1, and miR-153 can increase the expression level of BRCA1 (Fkih M'hamed et al. 2015). miR-146a and miR-146b-5p downregulate BRCA1 expression in TNBC (Garcia et al. 2011). However, it has also been reported that miR-146a expression positively correlates with BRCA1 expression and results in downregulation of EGFR in TNBC (Kumaraswamy et al. 2015). MicroRNAs have also been implicated in Epithelial to Mesenchymal transition (EMT) that drives metastasis in TNBC (Damiano et al. 2017). It was known previously that ZEB1 (also known as DeltaEF1), a transcription factor, downregulates the expression level of E-cadherin in TNBC (Eger et al. 2005). Damiano et al. (2017) reported that two serum miRNAs, miR-200c/miR-141 locus methylation result in downregulation of expression levels of miR-200c, favoring metastasis (Damiano et al. 2017). These two miRNAs have also been reported as potential biomarkers of ovarian cancer (Gao and Wu 2015). miR-200 family of miRNAs also play important roles in metastasis of TNBC. Humphries et al. (Humphries et al. 2014) showed that miR-200 family expressions are abnormally low in migratory TNBC cells. Moreover, overexpressing a single member of this family, miR-200b, reduces TNBC metastasis drastically in the mouse model by inhibiting the protein kinase C-alpha (Humphries et al. 2014). The EPHA2 oncogene encodes a tyrosine kinase that is a cell membrane receptor. miR-200b directly represses the translation of EPHA2 and prevents metastasis in TNBC (Tsouko et al. 2015).

3.6 Epigenetic Landscape of Ductal Carcinoma In Situ

The progression of Ductal Carcinoma in Situ (DCIS) to Invasive Ductal Carcinoma (IDC) occurs only in a relatively small percentage of patients and many patients with DCIS undergo surgery unnecessarily. Hence it is an important clinical challenge to identify the transition from DCIS to IDC (DeVaux and Herschkowitz 2018). We have observed in Chap. 2 that though the genomic landscapes of DCIS and IDC have differences, they are very similar and it is hard to implicate any particular genetic variation in early stages of IDC. Hence it is important to identify possible early epigenetic cues for IDC among DCIS patients. Though hypomethylation is a general characteristic of genes in many cancers, many tumor suppressor genes have been observed to be hypermethylated in breast cancer, causing the progression of carcinogenesis. Two tumor suppressor genes, CACNA1A and MGMT, showed significant methylation in IDC compared to DCIS (DeVaux and Herschkowitz 2018). CACNA1A is a gene that codes for a protein responsible for making calcium channels on cell membranes that are necessary for cell-to-cell communication.

One of the important epigenetic cues in breast carcinogenesis is dense methylation of CpG islands in the promoter regions of tumor suppressor genes, as dense methylation silences gene transcription by preventing the binding of transcription factors. However, it is important to know when this hypermethylation starts. Verschuur-Maes et al. (2012) investigated hypermethylation in the precursor columnar cell lesion (CCL). They observed increasing methylation from normal cells to CCL to DCIS to invasive carcinoma. The methylation levels were similar in DCIS and invasive carcinoma. The methylation levels of CCND2, CDH13, and ID4 increased significantly from normal breast cells to CCL to DCIS. On the other hand, the genes SCGB3A1, SFRP5, and RASSF1 were already significantly methylated in CCL, indicating that promoter hypermethylation of tumor suppressor genes is an early event in carcinogenesis (Verschuur-Maes et al. 2012). The SCGB3A1 gene is mostly expressed in the epithelium of the airways and has immunoregulatory functions (Reynolds et al. 2002). SFRP5 has important role as an inhibitor in the Wnt pathway (Cho et al. 2018).

Johnson et al. (2015) studied the differential DNA methylation of ductal carcinoma in situ (DCIS) cells and adjacent normal cells. They identified over 600 CpG loci in the promoter regions that were differentially methylated and associated with the progression of cancer. Most of these loci were associated with homeobox genes that control development. The association between hypermethylation of CpG islands in the promoter regions of homeobox genes and progression of cancer has been reported earlier (Fackler et al. 2011; Tommasi et al. 2009). The most hypermethylated genes in the study of Johnson et al. (2015) were TBX15, DLX4, EN1, and HOXB13. It has been reported that TBX15 hypermethylation is a poor prognosis in prostate cancer (Kron et al. 2012). DLX4 acts as a tumor suppressor gene and reduces invasion of cancer cells and metastasis (Tomida et al. 2007). HOXB13 has been implicated in tumor progression in breast cancer (Shah et al. 2013; Rodriguez et al. 2008).

3.7 Epigenetic Landscape of Invasive Lobular Carcinoma

Research on epigenetics of invasive lobular carcinoma (ILC) is still at an early stage, and most of the work has been concentrated on understanding altered methylation of CpG islands in the promoter regions of genes. Recently, Suman et al. (2021) studied genome wide alteration of DNA methylation patterns in ILC tumors. Their aim was to study the correlation between alteration of methylation and survival rates of patients with ILC. They first identified ten genes with most altered methylation patterns and further associated four genes with the highest correlation with survival rate. These four genes are HCG4P3, TMEM101, CELF2 and APC (Suman et al. 2021). TMEM101 is a gene responsible for activating the NF-κB signalling pathway (Matsuda et al. 2003). CELF2 is a RNA-binding protein that has an important role in T cell signalling-induced alternate polyadenylation for many proteins (Chatrikhi et al. 2019). APC interacts with E-cadherin and has been implicated in colorectal cancer (Markowitz and Bertagnolli 2009).

3.8 Conclusion

Epigenetics plays very important roles in breast cancer initiation, progression, and metastasis. Though genetic mutations are irreversible, aberrant epigenetic alterations can be targeted using drugs, a main aim of understanding epigenetic modulations in any cancer, and particularly in breast cancer. We have discussed the current understanding of epigenetics of breast cancer in this chapter. We have discussed the main epigenetic effects like DNA methylation, histone modifications, and miRNA silencing. We have then discussed the current understanding of epigenetic modifications for different subtypes of breast cancer. However, research on epigenetics of breast cancer is continuing and improved understanding in this area will improve diagnosis and drug development for different types of breast cancer. We will discuss in the next chapter the drugs that have been developed already with the help of current understanding of epigenetics of breast cancer. We have not discussed an important group of epigenetic modulations, post-translational modifications of proteins. There are many such post-translational modifications including acetylation, citrullination, methylation, glycosylation, sumoylation, ubiquitination, and palmitoylation. However, the main focus of research in epigenetics of breast cancer has been the epigenetic modulations we have discussed in this chapter, and research on the effects of post-translational modifications of proteins on breast cancer has started only recently. We refer to the review paper by Liu et al. (2022) for an overview of the current research in this area.

References

Alberts B et al (2015) Molecular biology of the cell, 6th edn. Garland Science, New York

Anfossi S et al (2014) High serum miR-19a levels are associated with inflammatory breast cancer and are predictive of favorable clinical outcome in patients with metastatic HER2þ inflammatory breast cancer. PLoS One 9:e83113. https://doi.org/10.1371/journal.pone.0083113

Annunziato A (2008) DNA packaging: nucleosomes and chromatin. Nature Education 1(1):26

Arderiu G et al (2007) HoxA5 stabilizes adherens junctions via increased Akt1. Cell Adhes Migr 1(4):185–195. https://doi.org/10.4161/cam.1.4.5448

Bailey ST, Westerling T, Brown M (2015) Loss of estrogen-regulated microRNA expression increases HER2 signaling and is prognostic of poor outcome in luminal breast cancer. Cancer Res 75:436–e445

Baugh EH, Ke H, Levine AJ, Bonneau RA, Chan CS (2017) Why are there hotspot mutations in the TP53 gene in human cancers? Cell Death Differ 25:154–160. https://doi.org/10.1038/cdd.2017.180

Baylin SB, Ohm JE (2006) Epigenetic gene silencing in cancer–a mechanism for early oncogenic pathway addiction? Nat Rev Cancer 6:107–116. https://doi.org/10.1038/nrc1799

Branham MT et al (2012) Methylation profile of triple-negative breast carcinomas. Oncogenesis 1: e17

Castilla MA et al (2012) MicroRNA-200 family modulation in distinct breast cancer phenotypes. PLoS One 7:e47709. https://doi.org/10.1371/journal.pone.0047709

Chan QK, Ngan HY, Ip PP, Liu VW, Xue WC, Cheung AN (2009) Tumor suppressor effect of follistatin-like 1 in ovarian and endometrial carcinogenesis: a differential expression and functional analysis. Carcinogenesis 30:114–121

Chatrikhi R, Mallory MJ, Gazzara MR, Agosto LM, Zhu WS, Litterman AJ, Mark Ansel K, Lynch KW (2019) Alternative polyadenylation by competing with enhancers of the polyadenylation machinery. Cell Rep 28:2795–2806. https://doi.org/10.1016/j.celrep.2019.08.022

Cho YK, Kang YM, Lee SE, Lee YL, Seol SM, Lee WJ, Park JY, Jung CH (2018) Effect of SFRP5 (Secreted Frizzled–Related Protein 5) on the WNT5A (Wingless-Type Family Member 5A)-induced endothelial dysfunction and its relevance with arterial stiffness in human subjects. Arterioscler Thromb Vasc Biol 38:1358–1367

Damiano V et al (2017) Epigenetic silencing of miR-200c in breast cancer is associated with aggressiveness and is modulated by ZEB1. Genes Chromosomes Cancer 56:147–158

Deaton AM, Bird A (2011) CpG islands and the regulation of transcription. Genes Dev 25(10): 1010–1022. https://doi.org/10.1101/gad.2037511

DeVaux RS, Herschkowitz J (2018) Beyond DNA: the role of epigenetics in the premalignant progression of breast cancer. J Mammary Gland Biol Neoplasia 23:223–235. https://doi.org/10.1007/s10911-018-9414-2

Dianatpur A et al (2018) Expression analysis of AFAP1-AS1 and AFAP1 in breast cancer. Cancer Biomark 22(1):49–54. https://doi.org/10.3233/CBM-170831

Eger A et al (2005) DeltaEF1 is a transcriptional repressor of E-cadherin and regulates epithelial plasticity in breast cancer cells. Oncogene 24(14):2375–2385. https://doi.org/10.1038/sj.onc.1208429

Elsheikh SE et al (2009) Global histone modifications in breast cancer correlate with tumor phenotypes, prognostic factors, and patient outcome. Cancer Res 69:3802–3809

Fackler MJ, Umbricht CB, Williams D, Argani P, Cruz LA, Merino VF et al (2011) Genome-wide methylation analysis identifies genes specific to breast cancer hormone receptor status and risk of recurrence. Cancer Res 71(19):6195–6207. https://doi.org/10.1158/0008-5472.CAN-11-1630

Falahi F, Huisman C, Kazemier HG, van der Vlies P, Kok K, Hospers GA, Rots MG (2013) Towards sustained silencing of HER2/neu in cancer by epigenetic editing. Mol Cancer Res 11: 1029–e1039

Fkih M'hamed I, Privat M, Ponelle F, Penault-Llorca F, Kenani A, Bignon YJ (2015) Identification of miR-10b, miR-26a, miR-146a and miR-153 as potential triple-negative breast cancer biomarkers. Cell Oncol (Dordr) 38:433–442

Fraga MF et al (2005) Loss of acetylation at Lys 16 and trimethylation at Lys20 of histone H4 is a common hallmark of human cancer. Nat Genet 37:391–400. https://doi.org/10.1038/ng1531

Gao YC, Wu J (2015) MicroRNA-200c and microRNA-141 as potential diagnostic and prognostic biomarkers for ovarian cancer. Tumor Biol 36(6):4843–4850. https://doi.org/10.1007/s13277-015-3138-3

Gao Y et al (2015) The integrative epigenomic-transcriptomic landscape of ER positive breast cancer. Clin Epigenetics 7:126. https://doi.org/10.1186/s13148-015-0159-0

Garcia AI et al (2011) Down-regulation of BRCA1 expression by miR-146a and miR-146b-5p in triple negative sporadic breast cancers. EMBO Mol Med 3:279–290

Gregoire JM et al (2016) Identification of epigenetic factors regulating the mesenchyme to epithelium transition by RNA interference screening in breast cancer cells. BMC Cancer 16:700

Grimson A, Farh KK, Johnston WK, Garrett-Engele P, Lim LP, Bartel DP (2007) MicroRNA targeting specificity in mammals: determinants beyond seed pairing. Mol Cell 27:91–105

Harburg GC, Hinck L (2011) Navigating breast cancer: axon guidance molecules as breast cancer tumor suppressors and oncogenes. J Mammary Gland Biol Neoplasia 16:257–270

Holm K, Hegardt C, Staaf J, Vallon-Christersson J, Jönsson G, Olsson H et al (2010) Molecular subtypes of breast cancer are associated with character-istic DNA methylation patterns. Breast Cancer Res 12:R36

Hu F et al (2013) BMP-6 inhibits cell proliferation by targeting microRNA-192 in breast cancer. Biochim Biophys Acta 1832:2379–2390

Hughes LA et al (2013) The CpG Island methylator phenotype: what's in a name? Cancer Res 73: 5858–5868

Humphries B et al (2014) MicroRNA-200b targets protein kinase Cα and suppresses triple-negative breast cancer metastasis. Carcinogenesis 35:2254–2263

International Human Genome Sequencing Consortium (2001) Initial sequencing and analysis of the human genome. Nature 409(6822):860–921. https://doi.org/10.1038/35057062

Jenuwein T (2006) The epigenetic magic of histone lysine methylation. FEBS J 273(14): 3121–3135. https://doi.org/10.1111/j.1742-4658.2006.05343.x

Johnson KC et al (2015) DNA methylation in ductal carcinoma in situ related with future development of invasive breast cancer. Clin Epigenetics 7:75. https://doi.org/10.1186/s13148-015-0094-0

Jones PA, Baylin SB (2007) The epigenomics of cancer. Cell 128(4):683–692. https://doi.org/10.1016/j.cell.2007.01.029

Jung EJ et al (2012) Plasma microRNA 210 levels correlate with sensitivity to trastuzumab and tumor presence in breast cancer patients. Cancer 118:2603–e2614

Kagara N et al (2012) Epigenetic regulation of cancer stem cell genes in triple-negative breast cancer. Am J Pathol 181:257–267

Kron K, Liu L, Trudel D, Pethe V, Trachtenberg J, Fleshner N et al (2012) Correlation of ERG expression and DNA methylation biomarkers with adverse clinicopathologic features of prostate cancer. Clin Cancer Res 18(10):2896–2904. https://doi.org/10.1158/1078-0432.CCR-11-2901

Kumaraswamy E et al (2015) BRCA1 regulation of epidermal growth factor receptor (EGFR) expression in human breast cancer cells involves microRNA-146a and is critical for its tumor suppressor function. Oncogene 34:4333–4346

Lee JA, Lee HY, Lee ES, Kim I, Bae JW (2011) Prognostic implications of microRNA-21 overexpression in invasive ductal carcinomas of the breast. J Breast Cancer 14:269–e275

Li Y et al (2014) Comparative epigenetic analyses reveal distinct patterns of oncogenic pathways activation in breast cancer subtypes. Hum Mol Genet 23:5378–e5393

Lindqvist BM, Farcas SA, Wingren S, Nilsson TK (2012) DNA methylation pattern of the SLC25A43 gene in breast cancer. Epigenetics 7(3):300. https://doi.org/10.4161/epi.7.3.19064

Lindqvist BM, Wingren S, Motlagh PB, Nilsson TK (2014) Whole genome DNA methylation signature of HER2-positive breast cancer. Epigenetics 9(8):1149–1162. https://doi.org/10.4161/epi.29632

Ling L, Maguire JJ, Davenport AP (2013) Endothelin-2, the forgotten isoform: emerging role in the cardiovascular system, ovarian development, immunology and cancer. Br J Pharmacol 168(2): 283–295. https://doi.org/10.1111/j.1476-5381.2011.01786.x

Lips EH et al (2013) Triple-negative breast cancer: BRCAness and concordance of clinical features with BRCA1-mutation carriers. Br J Cancer 108:2172–2177

Liu J, Wang Q, Kang Y, Xu S, Pang D (2022) Unconventional protein post-translational modifications: the helmsmen in breast cancer. Cell Biosci 12(22):22. https://doi.org/10.1186/s13578-022-00756-z

Markowitz SD, Bertagnolli MM (2009) Molecular origins of cancer: molecular basis of colorectal cancer. N Engl J Med 361(25):2449–2460. https://doi.org/10.1056/NEJMra0804588

Mathe A et al (2016) DNA methylation profile of triple negative breast cancer-specific genes comparing lymph node positive patients to lymph node negative patients. Sci Rep 6:33435

Matsuda A, Suzuki Y, Honda G, Muramatsu S, Matsuzaki O, Nagano Y et al (2003) Large-scale identification and characterization of human genes that activate NF-κB and MAPK signalling pathways. Oncogene 22(21):3307–3318. https://doi.org/10.1038/sj.onc.1206406

Mishra SK, Mandal M, Mazumdar A, Kumar R (2001) Dynamic chromatin remodeling on the HER2 promoter in human breast cancer cells. FEBS Lett 507:88–e94. https://doi.org/10.1016/s0014-5793(01)02951-9

Mungamuri SK, Murk W, Grumolato L, Bernstein E, Aaronson SA (2013) Chromatin modifications sequentially enhance ErbB2 expression in ErbB2 positive breast cancers. Cell Rep 5(2):302–313. https://doi.org/10.1016/j.celrep.2013.09.009

Musselman CA, Kutateladze TG (2011) Handpicking epigenetic marks with PHD fingers. Nucleic Acids Res 39(21):9061–9071

Nassar FJ et al (2014) miRNA as potential biomarkers of breast cancer in the Lebanese population and in young women: a pilot study. PLoS One 9:e107566. https://doi.org/10.1371/journal.pone.0107566

Ottaviano YL et al (1994) Methylation of the estrogen receptor gene CpG Island marks loss of estrogen receptor expression in human breast cancer cells. Cancer Res 54:2552–2555

Pathiraja T, Stearns V, Oesterreich S (2010) Epigenetic regulation in estrogen receptor positive breast cancer–role in treatment response. J Mammary Gland Biol Neoplasia 15(1):35–47. https://doi.org/10.1007/s10911-010-9166-0

Persson H et al (2011) Identification of new microRNAs in paired normal and tumor breast tissue suggests a dual role for the ERBB2/Her2 gene. Cancer Res 71:78–e86

Pfister S et al (2008) The histone acetyltransferase hMOF is frequently downregulated in primary breast carcinoma and medulloblastoma and constitutes a biomarker for clinical outcome in medulloblastoma. Int J Cancer 122:1207–1213

Radpour R et al (2009) Methylation profiles of 22 candidate genes in breast cancer using high-throughput MALDI-TOF mass array. Oncogene 28:2969–2978. https://doi.org/10.1038/onc.2009.149

Reynolds SD, Reynolds PR, Pryhuber GS, Finder JD, Stripp BR (2002) Secretoglobins SCGB3A1 and SCGB3A2 define secretory cell subsets in mouse and human airways. Am J Respir Crit Care Med 166(11):1498–1509. https://doi.org/10.1164/rccm.200204-285OC

Reynolds PA et al (2006) Tumor suppressor p16INK4A regulates polycomb-mediated DNA hypermethylation in human mammary epithelial cells. J Biol Chem 281:24790–24802

Rodriguez BA, Cheng AS, Yan PS, Potter D, Agosto-Perez FJ, Shapiro CL et al (2008) Epigenetic repression of the estrogen-regulated Homeobox B13 gene in breast cancer. Carcinogenesis 29(7):1459–1465. https://doi.org/10.1093/carcin/bgn115

Roll JD, Rivenbark AG, Jones WD, Coleman WB (2008) DNMT3b overexpression contributes to a hypermethylator phenotype in human breast cancer cell lines. Mol Cancer 7:15

Rossetto D, Avvakumov N, Cote J (2012) Histone phosphorylation. Epigenetics 7(10):1098–1108

Rountree MR, Bachman KE, Baylin SB (2000) DNMT1 binds HDAC2 and a new co-repressor, DMAP1, to form a complex at replication foci. Nat Genet 25:269–277

Shah N, Jin K, Cruz LA, Park S, Sadik H, Cho S et al (2013) HOXB13 mediatestamoxifen resistance and invasiveness in human breast cancer by suppressing ERalpha and inducing IL-6 expression. Cancer Res 73(17):5449–5458. https://doi.org/10.1158/0008-5472.CAN-13-1178

Shu S et al (2016) Response and resistance to BET bromodomain inhibitors in triple-negative breast cancer. Nature 529:413–417

Singla H et al (2017) Recent advances in HER2 positive breast cancer epigenetics: susceptibility and therapeutic strategies. Eur J Med Chem 142:316–327. https://doi.org/10.1016/j.ejmech.2017.07.075

Sogon T, Masamura S, Hayashi S, Santen RJ, Nakachi K, Eguchi H (2007) Demethylation of promoter C region of estrogen receptor alpha gene is correlated with its enhanced expression in estrogen-ablation resistant MCF-7 cells. J Steroid Biochem Mol Biol 105:106–114

Sossey-Alaoui K, Bialkowska K, Plow EF (2009) The miR200 family of microRNAs regulates WAVE3-dependent cancer cell invasion. J Biol Chem 284:e33019–e33029

Stirzaker C et al (2015) Methylome sequencing in triple-negative breast cancer reveals distinct methylation clusters with prognostic value. Nat Commun 6:5899. https://doi.org/10.1038/ncomms6899

Suman M, Dugue PA, Wong EM, Joo JE et al (2021) "Association of variably methylated tumour DNA regions with overall survival for invasive lobular breast cancer", clinical. Epigenetics 13(11):11. https://doi.org/10.1186/s13148-020-00975-6

Temian DC, Pop LA, Irimie AI, Berindan-Neagoe I (2018) The epigenetics of triple-negative and basal-like breast cancer: current knowledge. J Breast Cancer 21(3):233–243. https://doi.org/10.4048/jbc.2018.21.e41

Terada K et al (2009) Association between frequent CpG Island methylation and HER2 amplification in human breast cancers. Carcinogenesis 30:466–471. https://doi.org/10.1093/carcin/bgp021

Tomida S, Yanagisawa K, Koshikawa K, Yatabe Y, Mitsudomi T, Osada H et al (2007) Identification of a metastasis signature and the DLX4 homeobox protein as a regulator of metastasis by combined transcriptome approach. Oncogene 26(31):4600–4608. https://doi.org/10.1038/sj.onc.1210242

Tommasi S, Karm DL, Wu X, Yen Y, Pfeifer GP (2009) Methylation of homeobox genes is a frequent and early epigenetic event in breast cancer. Breast Cancer Res 11(1):R14. https://doi.org/10.1186/bcr2233

Tsouko E, Wang J, Frigo DE, Aydoğdu E, Williams C (2015) miR-200a inhibits migration of triple-negative breast cancer cells through direct repression of the EPHA2 oncogene. Carcinogenesis 36:1051–1060

Veeck J et al (2008) Wnt signalling in human breast cancer: expression of the putative Wnt inhibitor Dickkopf-3 (DKK3) is frequently suppressed by promoter hypermethylation in mammary tumours. Breast Cancer Res 10:R82

Verschuur-Maes AH, de Bruin PC, van Diest PJ (2012) Epigenetic pro-gression of columnar cell lesions of the breast to invasive breast cancer. Breast Cancer Res Treat 136(3):705–715

Wee E et al (2012) Mapping the regulatory sequences controlling 93 breast cancer-associated miRNA genes leads to the identification of two functional pro-moters of the Hsa-mir-200b cluster, methylation of which is associated with metastasis or hormone receptor status in advanced breast cancer. Oncogene 31:4182–e4195

Wiesmann F et al (2009) Frequent loss of endothelin-3 (EDN3) expression due to epigenetic inactivation in human breast cancer. Breast Cancer Res 11(3):R34

Xi Y et al (2018) Histone modification profiling in breast cancer cell lines highlights commonalities and differences among subtypes. BMC Genomics 19:150

Yamaguchi T, Mukai H, Yamashita S, Fujii S, Ushijima T (2015) Comprehensive DNA methylation and extensive mutation analyses of HER2-positive breast cancer. Oncology 88:377–e384

Breast Cancer Therapy and Control

4

4.1 Introduction

The purpose of any cancer treatment is to eradicate the offending cancerous growth, so as to prevent the recurrence or spread of primary cancer. This is also the goal for breast cancer therapy. The main objective of a good therapeutic approach is the extermination of the tumor from breast and adjoining lymph nodes, for preventing metastatic recurrence. For tumors which are not metastatic, surgical elimination of the growth and resection of lymph nodes are practised; often radiation is given after surgery. Neoadjuvant therapy is given prior to surgery, and adjuvant therapy is given after surgery. Both neoadjuvant and adjuvant therapies may be given depending on the health conditions of a patient. The progression of cancer is at a critical stage, and the prognosis is worse when cancer has metastasized. Oncologists usually emphasize on prolonging life span and palliation after metastasis, in order to provide a better quality life for the patient. Neoadjuvant and adjuvant therapies are practised in metastatic breast cancer. We will first discuss chemotherapy, the conventional therapy for destroying cancer cells using drugs. We will next discuss the pitfalls of conventional therapy due to adverse side effects of drugs. This will bring us to the use of natural compounds for breast cancer prevention and treatment and their advantages.

4.2 Drugs Targeting Genetics and Epigenetics of Breast Cancer

Nonsurgical breast cancer treatments aim to destroy cancer cells. These treatments include:

- Radiation
- Hormone therapies
- Chemotherapies

M. Roy et al., *Genetics and Epigenetics of Breast Cancer*,
https://doi.org/10.1007/978-981-19-9925-3_4

- Targeted therapies
- Immunotherapies (sometimes called biological therapies)

4.2.1 Genetic and Epigenetic Targets for Breast Cancer Drugs

Initiation of breast cancer and progression from the premalignant stage to the invasive stage are due to accumulation of genetic as well as epigenetic alterations. Epigenetic changes are responsible for altered gene activity, as well as gene expression. A substantial percentage of breast cancer has a hereditary history. Mutations in BRCA1 and BRCA2 are intricately associated with the etiology of breast cancer; these are high penetrance genes. In order to have an idea of risk of cancer, values of multigene panels are gaining importance. In genetics, penetrance is the proportion of individuals who carry a particular variant of a gene. The susceptibility of cancer with mutations in moderate-penetrance genes is lower, compared to those in high penetrance genes (Tung et al. 2016). Among the medium penetrance genes that are implicated in causation of breast cancer are ATM, CHEK2, PTEN, PALB2, TP53, and STK11 (Buocikova et al. 2020). The changes mentioned in this paragraph involve the activation or inactivation of several genes, oncogenes, and tumor suppressor genes, respectively. Mutations of oncogenes usually lead to gain of function, while mutations in tumor suppressor genes lead to loss of function, both of these are responsible for malignant transformations (Vogelstein and Kinzler 2004).

Tumor suppressor genes slow down cell division, take care of repairing DNA damage and copying mistakes, and control apoptosis, that is, guide a cell when to die. Deregulation of tumor suppressor genes leads to uncontrolled growth or proliferation of cells due to breakdown of the apoptosis pathway, leading to cancer. Inactivation of tumor suppressor genes may occur in three ways: firstly, through mutation, when they fail to function properly; secondly, loss of heterozygosity, leading to complete loss of a gene; thirdly switching off a gene due to epigenetic changes. Deregulation of epigenetic machinery may cause epigenetic silencing, for example, inappropriate methylation of cytosine residues in CpG sequences in promoter regions, which control gene expression (Mithraprabhu et al. 2014; Baylin and Jones 2016; Handy et al. 2011).

A correlation between genetic alterations, cancer initiation, its promotion and progression have been known for quite some time. Most of these alterations are due to a plethora of epigenetic changes, which include post-translational modifications, DNA methylation, and noncoding RNAs that affect transcriptional gene silencing (Barneh et al. 2019). It is understood now that epigenetics plays a vital role in the pathophysiology of breast cancer, no less than genetics. To develop targeted therapeutic approaches, unravelling the mechanisms by which specific genes cause epigenetic regulations is very important. Many of the epigenetic therapies are based on modulation of chromatin structure. For example, targeting histone deacetylase (HDAC) and histone demethylase (HDM) inhibitors may be a good therapeutic modality for the treatment of cancer. These epigenetic inhibitors can cause changes in DNA methylation or acetylation patterns of histones. They can hinder the recognition of acetylated lysine residues. These epigenetic inhibitors may be able to restore the epithelial phenotypes during epithelial to mesenchymal transition (EMT). Various drugs in conjunction with HDAC inhibitors may prevent EMT which plays a leading role in progression of cancer and finally metastasis (Hyun et al. 2017).

Expression of epigenetic proteins in diseased and normal condition is different. DNMTs (DNA methyl transferases) are responsible for C5-methylation of CpG dinucleotide sequences; a distinct shift of methylation pattern is noticed in case of cancer. Transcription is initiated in the promoter regions of genes, and methylation of DNA in the promoter regions results in inactivation or silencing of genes. Silencing of genes may result in halting of several vital pathways. One such is the DNA repair pathway, which aids in halting tumor cell proliferation. HDACs (Histone deacetylases) are epigenetic erasers (they remove acetyl groups), which are found to be overexpressed in cancers and are associated with repression of genes.

Histone modifications are epigenetic regulatory players, having roles in various cellular events. Lysine methylations occur on globular domains and tails of histones, and their levels are controlled by methyltransferases ("writers") and demethylases ("erasers"). Specialized proteins ("readers") recognize specific methyl-lysines depending on amino-acid sequence and methylation state. Mutations of methyltransferases, demethylases, and methyl-lysine-binding proteins have been linked to various human diseases (Rodríguez-Paredes and Esteller 2011). Hence histone methylation-related proteins may be good therapeutic targets. All three kinds of epigenetic proteins, namely readers, writers, and erasers may be employed as drug targets.

MicroRNAs (miRNA) may regulate gene expression, though their short size does not enable them to code for proteins. The miRNAs are small single stranded non-coding RNA molecules, containing about 22 nucleotides. They aid in RNA silencing and post-translational regulation of protein coding genes. Genes responsible for causation of cancer, particularly those involved in multidrug resistance may be regulated by miRNAs. Gene silencing by miRNA may be due to cleavage of the mRNA strand into two pieces; shortening of poly(A) tail, hindering the stability of mRNA; inefficient translation of the mRNA into proteins by ribosomes. Though small, these miRNAs play a leading role in development of chemosensitivity and chemoresistance in various types of cancer, including breast cancer. The miRNAs that are involved in drug resistance in breast cancer are miR-21, that targets PTEN, PDCD4, and may increase the sensitivity of tumors to chemotherapy agent trastuzumab (De Mattos-Arruda et al. 2015). miR-25 targets ABCG2 and enhances the sensitivity of breast tumors to epirubicin (Wang et al. 2014). miR-134 increases the sensitivity of tumor cells to doxorubicin, by targeting MRP1/ABCC1 (Lu et al. 2015). Another miRNA, miR-217 increases the efficiency of tumor cells to tamoxifen, etoposide, and lapatinib by targeting PTEN (Zhang et al. 2018). Therefore, these miRNAs may enhance the efficacy of breast tumor cells to chemotherapy agents and may thus overcome drug resistance. Besides, they are good prognostic biomarkers and also predict survival in response to chemotherapy (de Montes et al. 2015; Roessler et al. 2015).

miRNAs are considered as a pool of highly potent biomarkers. miRNA profiles may be used to classify breast tumors as luminal A and luminal B, basal, HER2 positive, etc. Also, abnormal expression of miRNAs has been found to have correlation with a number of clinical traits, like angiogenesis, metastasis, EMT, etc. Organized screening of epigenetic marks may suggest breast cancer biomarkers, which may help to improve the quality of treatment with good outcome, particularly in terms of precision medicine. Aim of the biomarker driven treatment approach is to identify the important biomarkers, targeting which may achieve good clinical outcomes. In order to implement a good therapeutic regimen, expression levels of the biomarkers, involvement of lymph node (if any), and determination of tumor size are very important. Nonhistone nuclear protein Ki-67, which is a cell proliferation marker, is important to consider, as this marker is highly expressed in malignant tissues and in cells which are poorly differentiated. Therefore, Ki-67 is a good measure of the aggressiveness of tumor. As we have discussed, there are a number of different therapeutic measures for non-metastatic breast cancer, including surgical elimination of the offending growth, radiation therapy or radiotherapy, adjuvant or neoadjuvant chemotherapy, hormonal or endocrine therapy, and targeted therapy (Chew 2001).

With the progression of cancer, some genetic anomalies develop, which are highly prone to developing resistance to cancer therapy. Epigenetic changes lead to resistance to drugs, these factors are suggestive of the personalized variation in response to chemotherapy. These modifications may be reverted employing epigenetic therapy. Hence, reprograming of epigenome implicated in cancer may be a target for cancer therapy, as well as for reversal of drug resistance. Therefore, development of drugs that may reprogram the epigenetic landscape has gained in importance in recent years.

Inhibitors of DNMT (5-Azacytidine or AZA, 5-Aza-2′-deoxycytidine or DAC) and HDAC (Vorinostat or VOR, phenylbutyrate) have gone into clinical trials. In breast cancer monotherapy, these drugs failed to show any promising result even at the highest tolerated dose. Hence, usage of these drugs was considered unsuitable. These epi-drugs had also been used in conjunction with other cytotoxic agents, radiation therapy, targeted therapy, hormone therapy, and immunotherapy. Combination of two epi-drugs was also tried clinically. 5-Aza-2′-deoxycytidine or DAC was extensively studied in preclinical trials in breast cancer. For the combinatorial therapy, outcome is better compared to monotherapy. However, due to systemic toxicities and poor efficacy, overall acceptance is poor. Drugs that inhibit HDAC have shown to inhibit the growth and proliferation of tumors having minimal toxicity, both in in vivo and in vitro models. In in vitro models, these inhibitors can inhibit proliferation of cells, induce apoptosis, inhibit invasion and migration, thereby preventing metastasis (Bian et al. 2018). miRNAs can be negative regulators of gene expression at both transcriptional and post-transcriptional levels. In breast cancer, aberrant expressions of miRNAs have been found. The miR-200 family is reported to be a suppressor of epithelial–mesenchymal transition (EMT) and invasion, hence an inhibitor of metastasis. In breast cancer, miR-200c serves as a tumor suppressor and regulates tumor angiogenesis, proliferation, invasion, and migration (Chen et al. 2012). There are a number of epi-drugs used in combination with other treatment modalities, but, not with much promising results. Therefore, identification of newer epigenetic targets, which will give promising outcomes is needed. Maximization of efficacy, besides minimization of side effects is the goal of a good therapeutic approach. HDACi have therapeutic effects against triple-negative breast cancer (TNBC) as well, such as sodium butyrate, vorinostat, panobinostat, mocetinostat, entinostat, etc. HDAC5 has been found to be overexpressed in breast cancer tissues. Therefore, this may be a good prognostic marker and a therapeutic agent. Besides, HDAC1 and HDAC3 are also highly expressed in breast cancer. HDAC1 inhibitors like vorinostat and entinostat may be promising therapeutic agents (Sher et al. 2022).

A combinatorial therapy of anticancer agents with inhibitors of HDAC may yield good therapeutic outcomes. LMK-235, a HDAC5 inhibitor in combination with an anticancer agent bortezomib is a novel combination in breast cancer therapy (Li et al. 2016). Another HDAC inhibitor is vorinostat, along with tamoxifen. The combination of endocrine therapy (tamoxifen) and HDAC inhibitor (vorinostat) showed significant reversal of hormone resistance in ER-positive advanced metastatic BC patients (Sher et al. 2022).

Some of the epi-drugs for breast cancer therapy which have been approved by FDA are DNMT inhibitors (azacitidine, decitabine, belinostat, fingolimod, panobinostat) and HDAC inhibitors (romidepsin, valproic acid, vorinostat). There are many other molecules which are under trial. The approach of reprogramming the cancer epigenome is a promising targeted therapy for treatment and reversibility of cancer drug resistance, thus enhancing chemotherapy response (Li et al. 2016).

Traditional epi-drug discovery is expensive and time-consuming (Naveja et al. 2016; Moreira-Silva et al. 2020). Hence, the focus has shifted to identifying novel potential epi-targets by already approved drugs through drug repositioning or repurposing (Blatt and Corey 2013; Shim and Liu 2014). Epi-drug repurposing has the advantage of exploring a wide diversity of molecular combinations in cancer, where a combination of epigenetic drugs is more effective than a single epi-drug that may result in resistance to chemotherapy (Moreira-Silva et al. 2020). To further understand this, we refer to the discussion in Chap. 3 on the mechanisms involved in epigenetic alterations in cancer and their relevance in cancer therapy.

4.2.2 Conventional Drugs Targeting Genetics and Epigenetics of Breast Cancer

Treatment of breast cancer is multifaceted, complex, and patient-specific. Among the interdisciplinary therapeutic modalities, correct regimen is decided by the stage and type of cancer. Therefore, proper diagnosis is the best guide in treatment management. Depending on the presence or absence of molecular markers implicated in breast cancer, it can be classified into three major subtypes. As discussed in previous chapters, these markers are estrogen receptor, progesterone receptor, and human epidermal growth factor 2 (HER2). The cancer subtypes can be classified as hormone receptor-positive, but HER2 negative; hormone receptor-positive with HER2 positive; all these markers are negative (triple negative). Some of the chemotherapeutic drugs used in the treatment of breast cancer, with their target, mode of action, and probable side effects have been listed in Table 4.1.

We have discussed in detail in this and previous chapters that genetic and epigenetic alterations may lead to aberrant gene expressions, culminating in breast carcinogenesis. Usually, a multidisciplinary approach is followed for treatment of breast cancer. With advancement of therapeutic modalities, efficacy of therapy has been enhanced, with a concomitant reduction is mortality rate, resulting in better outcomes (Moo et al. 2018). There are several pitfalls even after attaining a good prognosis. All the treatment regimens have their side effects. Besides, certain subtypes of breast tumor are difficult to control. Being a heterogeneous disease, breast cancer differs among different patients, which is termed inter-tumor heterogeneity. There is heterogeneity even within the same tumor, which is called intra-tumor heterogeneity (Turashvili and Brogi 2017). This warrants adoption of personalized therapy. Apart from all these complications, another important obstacle to overcome is incurable metastatic stage. Epigenetic alterations are often implicated in carcinogenesis. Some novel drugs have been found that target chromatin and other associated proteins (Verma and Kumar 2018). As we have discussed, use of epigenetic drugs (also called epi-drug) is gaining in importance. These drugs which are DNA methylation inhibitors and histone deacetylase inhibitors are used to treat specific cancers. Epi-drugs with conventional drugs may improve outcomes of therapeutics (Ganesan et al. 2019).

In the beginning, more than 90% of breast cancers are not metastatic. Hence, systemic therapy is followed in the early stages. Endocrine therapy is given to hormone receptor-positive patients. Chemotherapy is also used in some cases. For HER2 positive cancers, antibodies that target HER2 or small-molecule inhibitor therapy along with chemotherapeutic drugs are given. For triple-negative cancers, chemotherapy is given. Surgery, followed by radiation is practiced for non-metastatic breast cancer. For non-metastatic carcinoma, hormone receptor-positive, but HER2 negative breast tumors, typical systemic therapies are opted. In these cases, tamoxifen, letrozole, anastrozole, or exemestane are used as endocrine therapy.

There are a few combinations of chemotherapy used for breast cancer, these are (a) Adriamycin (generic name doxorubicin) and cyclophosphamide combination; (b) Adriamycin, cyclophosphamide, and paclitaxel combination; (c) Docetaxel and cyclophosphamide combination. In case of triple-negative breast cancer, similar chemotherapeutic regimens are followed. HER2 positive breast tumors may be hormone receptor-positive (HR+) or hormone receptor-negative (HR−), having distinct clinical behavior. HR−/HER2+ and HR+/HER2+ variants of breast cancer showed distinct histopathologic traits that may be related to their specific clinical behavior. Chemotherapy in conjunction with HER2 targeted therapy is given for all tumors, irrespective of hormone receptor status. These are (a) paclitaxel and

Table 4.1 Chemotherapeutic drugs used in breast cancer treatment, their targets, mode of action, and the side effects

Drugs used in breast cancer	Target	Mechanism of action	Side effects
Anthracyclines (doxorubicin and epirubicin)	DNA, topoisomerase II	Binds to DNA by intercalation between base pairs; interacts with DNA-topoisomerase II activity; base modification results, which in turn are responsible for disturbances in DNA replication and transcription.	Cardiotoxicity, alopecia, nausea, low blood count, mouth sore, shortness of breath
5-fluorouracil	Incorporates into DNA; T-RNA maturation	Blocks thymidylate synthase, preventing formation of dTMP, leading to "thymineless" death	Bruising and bleeding, anemia, mouth sore, loss of appetite, risk of infection
Carboplatin	DNA	Forms reactive platinum complexes that cause the intra- and inter-strand cross-linkage of DNA molecules within the cell	Black stool, blood in urine and stool, fever, cough, hoarseness, kidney toxicity, ototoxicity
Taxanes (paclitaxel and docetaxel)	Microtubules	Mitotic inhibitor; disrupts microtubule function, which are needed for cell division	Alopecia, arthralgia, myalgia, peripheral neuropathy nausea, vomiting
Cyclophosphamide (Cytoxan)	DNA	Cross-linking and alkylation of DNA that prevents synthesis of DNA and transcription	Alopecia, anorexia, nausea and vomiting, low blood count
Selective estrogen receptor modulators (SERM), raloxifene, tamoxifen	Estrogen receptor (ER)	Binds to ER, causing a conformational change leading to blockage of estrogen-dependent genes	Cardiotoxicity, breathing difficulty, hot flashes, vaginal bleeding, blood clots
Herceptin	Human epidermal growth factor receptor	Blocks intracellular signalling pathways. Binds to an extracellular domain and inhibits HER2 homodimerization, thereby preventing HER2-mediated signalling	Respiratory problem, cardiotoxicity, peripheral neuropathy, alopecia, flu-like syndrome
Bevacizumab	VEGF or its receptor	Selectively binds to circulating VEGF, inhibiting its binding to surface receptors. This limits the blood supply to tumor tissues	Upper respiratory infection, alopecia, abdominal pain, constipation, nausea, nose bleeding

trastuzumab; (b) Adriamycin, cyclophosphamide, paclitaxel, and trastuzumab; (c) docetaxel, carboplatin, and trastuzumab. Endocrine therapy may be given if the tumor is hormone receptor-positive. Tamoxifen, letrozole, anastrozole, or exemestane are generally used.

The main purpose of treatment is to get rid of the offending growth and proliferation to the nearby lymph nodes, so that metastatic relapse does not occur. To check metastatic relapse, systemic therapies comprising anthracycline and chemotherapy with taxane are given. Taxane-based chemotherapy is a popular treatment for breast cancer, and this prevents progression and overall survival. Docetaxel is a common taxane-based chemotherapy and can significantly improve progression-free survival as well as overall survival of patients. Hormonal therapy, like antiestrogen, anti-HER are common treatment methods for breast cancer. For metastatic breast cancer, growth spreads to bones, brain, liver, and lungs and are generally not curable. Treatments followed for metastatic breast cancer are chemotherapy, endocrine therapy, immunotherapy, and targeted therapy, but the choice of therapy depends on the tumor subtype, size of the tumor, if the growth has spread to other parts, i.e. whether it has metastasized (Royce and Osman 2015).

Analogues of rapamycin, everolimus along with exemestane have been tested clinically in postmenopausal breast cancer, where the growth is at an advanced stage. Everolimus has been given approval for use along with exemestane for HER2 negative postmenopausal breast cancer patients. This treatment is also given to hormone receptor-positive patients whose treatment with non-steroidal aromatase inhibitors does not succeed. Reports from a number of such trials had shown that analogues of rapamycin, in conjunction with everolimus may be a promising treatment modality for breast cancer (Hare and Harvey 2017). However, use of Rapalogues often activates feedback pathways that limit their usefulness. New types of inhibitors are currently under investigation. These inhibitors work through different mechanisms, for example, inhibitors that are ATP competitive with mTOR and which intervene in signalling from mTOR complexes. As such, the mammalian target of rapamycin or mTOR has important roles in breast cancer pathogenesis. It has been shown in several clinical trials that the mTOR inhibitor everolimus improves patient outcomes in human epidermal growth factor receptor–negative metastatic disease when it has progressed after endocrine therapy. Similar beneficial effects have been observed in hormone receptor–positive breast cancer as well. Inhibition of mTOR slows tumor growth and limits the rate of spread of cancer. Rapalogues are important in inhibiting the mTOR pathway. HER-2 negative postmenopausal patients are treated with a combination of everolimus (RAD-001/Afinitor), a rapalogue, and exemestane.

The serine/threonine kinase belongs to the phosphoinositide 3-kinase related kinase (PIKK) family is mTOR, which is highly expressed in many cancers, including breast cancer. Full form of mTOR is mammalian target of rapamycin. Rapamycin, along with its analogues inhibits mTOR signalling and slows down growth of tumors and its spread to other parts. There had been a number of clinical trials, where everolimus, a mTOR inhibitor has shown its potential to improve prognosis in breast cancer.

In order to find epigenetic inhibitors, first generation of epi-drugs was invented. These epi-drugs are mainly DNMTi and HDACi. DNMTi form covalent DNA adducts that cause DNA damage response activation. Hence as a treatment, DNMTi are incorporated into DNA during replication, and this prevents proliferation of aberrant DNA, eventually leading to apoptosis. However, there are cytotoxic implications in this treatment (Vogelstein and Kinzler 2004; Royce and Osman 2015; Liu et al. 2020a). HDACi are molecules that inhibit HDAC enzymes that are Zn^{2+} dependent, except sirtuin inhibitors, which inhibit NAD+ dependent histone deacetylases (Morel et al. 2020). However, first-generation HDACi had poor bioavailability and target selectivity. These drugs showed better efficacy in a pH that does not match the physiological value, also they show short half-life for degradation. In order to cope with this, second-generation epi-drugs had been created. These included DNMTi (such as guadecitabine and zebularine) and HDACi (including belinostat and panobinostat, hydroxamic acid, valproic acid, and tucidinostat), but, this time with better pharmacological properties. The second-generation drugs were designed in such a way that they had better inhibitory action with less side effects. The **third generation of epi-drugs** were developed with better understanding of epigenetic interactions. Hence, in order to develop highly efficient epi-drugs, better understanding of the whole set of interactions of epigenetic proteins in a cell is essential. Third generation epi-drugs include histone demethylase inhibitors (HDMi), histone methyltransferase inhibitors (HMTi), and bromodomain and extra-terminal domain inhibitors (BETi) (Montalvo-Casimiro et al. 2020).

A number of epi-drugs have received approval from Food and Drug Administration (FDA) for the treatment of cancers: 5-Azacytidine, 5-Aza-2′-deoxycytidine, vorinostat, romidepsin, belinostat, panobinostat, chidamide. Of these, 5-Azacytidine and 5-Aza-2'deoxycytidine are nucleoside analogs and are DNA methyl transferase inhibitors. Suberoylanilide hydroxamic acid (SAHA), vorinostat, trichostatin A (TSA) are hydroxamic acids and are histone deacetylase inhibitors. Cyclic peptides like trapoxin A and romidepsin are also histone deacetylase inhibitors. All these are first-generation epi-drugs, whose efficiency was not promising. However, modulation of epigenetic factors may have a vital role in cancer management. Among the second-generation epi-drugs that went for trial are some nucleoside analogs like zebularine; some non-nucleoside analogs like hydralazine and procainamide; some bi-substrate analogs like quinolone-quinazoline conjugates. All these are DNA methyl transferase inhibitors. Other second-generation epi-drugs are histone deacetylase inhibitors, which are hydroxamic acids like belinostat, dacinostat, panobinostat, abexinostat, ricolinostat, quisinostat, tefinostat; some are benzamides like tacedinaline, entinostat, mocetinostat, chidamide. Some carboxylic acids which are also histone deacetylase inhibitors belong to second-generation epi-drugs; some of these drugs are sodium butyrate, phenyl butyrate, valproic acid, and citarinostat.

Aberrant DNA methylation has been documented in breast cancer (Portela and Esteller 2010). Once hypomethylation is blocked, chromosomal instability and altered gene expression are inevitable. Promoter regions of tumor suppressor genes have been found to be hypermethylated in breast cancer. Also, there are a plethora of genes involved in regulation of cell cycle, repair of DNA, cellular adherence and metastasis, cell growth and homeostasis, metabolism and programmed cell death or apoptosis, and genes that code for epigenetic enzymes have been found to be hypermethylated in breast cancer (Stefansson and Esteller 2013). Anomalous hypermethylation of some key genes implicated in breast cancer may be good diagnostic, as well as prognostic markers. A link between tamoxifen resistance and aberrant hypermethylation of homeodomain transcription factor-2 ((PITX2) has been found (Stefansson and Esteller 2013). Genes that code for ER and PR are often hypermethylated, leading to development of ER and PR negative breast carcinoma. These events are reversible and this very nature made the epigenetic alterations attractive targets for drug development. Inhibitors of DNMTs and HDACs therefore are gaining momentum in breast cancer therapy. Inhibitors of DNMT aid in preventing re-methylation of DNA after cell division; these are classified as analogues of nucleoside and non-nucleoside. Nucleoside analogues are azacitidine and decitabine, which block the function by trapping them through formation of covalent bonds, when incorporated into DNA (Locke and Clark 2012). Of these two, azacitidine may be incorporated in both DNA and RNA. A link between reduction of size of breast tumor and intervention of therapeutic doses of azacitidine has been found (Falahi et al. 2014). Decitabine reactivates silenced genes by prevention of DNA re-methylation. In triple-negative breast cancer, decitabine can induce TNF-related apoptosis. In this way, inhibitors of DNMT which are nucleoside analogs can sensitize the breast cancer cells to chemotherapy drugs (Borges et al. 2013).

Metastatic breast cancer is often incurable, and the main aim of any therapy is to prolong life. Usually some combinations of targeted therapy, immunotherapy, chemotherapy, and endocrine therapy are treatment options in metastatic breast cancer. A choice of therapy is based on the spread of the disease, tumor subtype, and the presence of any molecular alterations. The effective new biological therapies include mTOR inhibitors (Li et al. 2020). However, more predictive biomarkers are needed for these treatments.

4.3 Problems with Conventional Therapy

Treatment protocols affect cancer cells as well as normal/healthy cells. This leads to a number of side effects/adverse effects. Common side effects are anorexia, i.e., loss of appetite; nausea and emesis, i.e., vomiting; weakness and lethargy; soreness in the mouth; alopecia, i.e., hair loss; obesity, i.e., weight gain; proneness to infections; bleeding and hemorrhage; diarrhea; early or premature menopause. Besides, there may be infertility, chemo fog or brain fog, emotional impact, etc. Patients getting chemotherapy often become irritable, worried, sad, dull, indifferent, and anxious. Psychological counseling may be of help in these situations. A number of side effects of the therapies used to treat breast cancer has been depicted in Fig. 4.1. However, there are many other pitfalls of therapy.

All patients do not have similar side effects, responses may be different for each individual, though the line of treatment may be similar. Patients undergoing

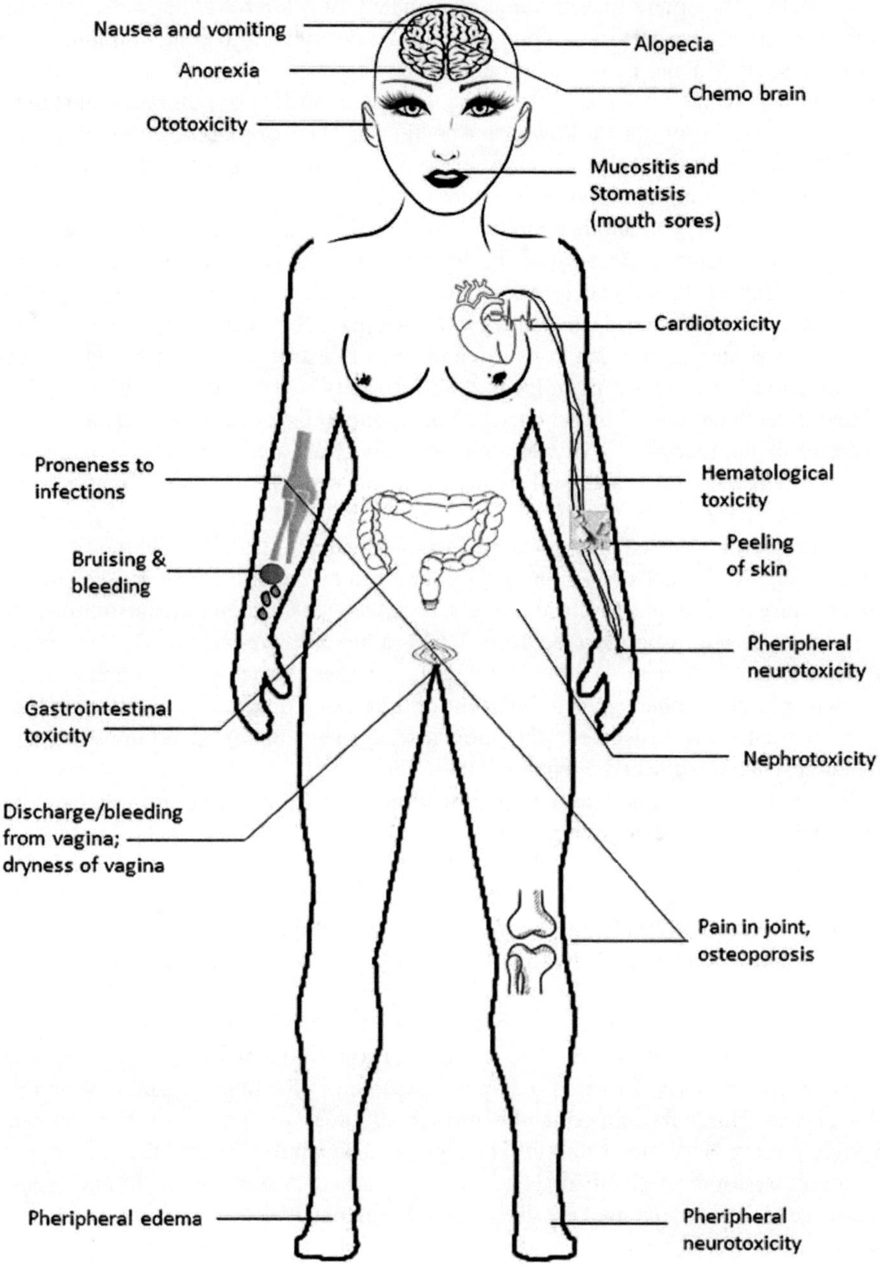

Fig. 4.1 Side effects occur when healthy tissues or organs get affected by cancer treatments, which are meant to kill cancer cells. Some of the side effects on the body have been depicted

chemotherapy treatment may become constipated, or often have diarrhea, irregular menstruation, early onset of menopause, tingling or prickling feeling, numbness, or paralytic sensation and pain. Radiotherapy may lead to swelling, or massive edema (also known as anasarca), pain, burning sensation, blemish or discoloration of skin at the site, blistering of skin followed by peeling of skin. Sometimes, though infrequently, radiation may burn the lung, leading to pneumonitis. However, generally swelling of the pulmonary tissues goes away with time.

Hormone therapy is another mode of treatment for breast cancer. The hormone estrogen can increase the growth of breast cancer cells. Certain drugs prevent estrogen to attach to the cancer cells. These drugs cause hot flashes, discharge, or bleeding from vagina and also dryness of vagina. One such drug is tamoxifen. Aromatase is an enzyme that converts androgen to estrogen. Therefore, inhibitors of aromatase lower the estrogen level. Anastrozole is an aromatase inhibitor, which is used in the treatment of breast cancer. This group of drugs may cause pain in joint and muscle, hot flashes, and dryness of the vagina. They also affect bones, i.e., raise the risk of osteoporosis and broken bones. Cholesterol levels may sometimes be enhanced.

Anticancer drug resistance developed during treatment is due to alteration in the drug targets. All the cells constituting a tumor are not identical. This heterogeneity in tumor causes resistance to the treatment regimen, where a significant number of tumor cells do not respond to the drug. This is a big problem from therapeutic point of view (Mansoori et al. 2017). There are a number of factors that contribute to resistance to chemotherapy, e.g., evasion of apoptosis, drug inactivation, development of multi-drug resistance, alteration in drug metabolism, gene amplification, increased DNA repair, and epigenetic alterations.

Immunotherapy is another new treatment modality. This is also not free from its side effects. Immunotherapy may cause fatigue, fever, chills, body aches and pains, sore throat, running nose, headache, etc. Besides these flu-like symptoms, immunotherapy may cause gastrointestinal problems, and normal liver function may get affected. Respiratory problems may develop, like cough and shortness of breath. Most of these side effects are not too severe and can be reversed if managed properly.

Targeted therapy is another type of treatment that targets proteins involved in the division, growth, and invasion of cancer cells. Targeted therapy is based on the idea of precision medicine. Targeted therapy is capable of blocking the action of abnormal protein Her2 that supports the growth of breast cancer cells. For patients overexpressing Her2, trastuzumab (Herceptin) or lapatinib (Tykerb) are prescribed. Of these, lapatinib is given for Her2 positive cells and used for metastatic breast cancer cells, which have metastasized to the brain as this drug can cross the blood–brain barrier. Trastuzumab is a drug which is a monoclonal antibody that targets Her2 positive cells. Trastuzumab is given intravenously. There are several side effects, first, intervention of this drug may cause chill and fever. Other side effects include nausea and vomiting, lethargy and weakness, diarrhea, breathlessness, headache, and rashes. This drug may damage the heart, leading to heart failure. Lapatinib, on the other hand, is given orally. Side effects are similar to those caused by trastuzumab. Another drug used in targeted therapy is bevacizumab. This drug aims to prevent formation of blood vessels that feed the tumor.

Palbociclib, a kinase inhibitor, is used to treat a certain type of hormone receptor-positive, advanced breast cancer or metastasized breast cancer. This drug blocks the action of the protein that helps cancer cells to multiply, thereby checking the spread of the disease. Some serious side effects include breathing difficulty, arrythmia, weakness, unusual bleeding, even from the nose.

Treatment with azacitidine and decitabine shows some toxic effects including gastrointestinal discomfort and myelosuppression. In case of romidepsin and vorinostat, side effects include fatigue, along with gastrointestinal discomfort and myelosuppression. Usage of epi-drugs in breast cancer showed some potential though they have not proved efficacy as mono therapy; however, pre-treatment and post-treatment biopsies may cause significant decrease in the DNA methylation after monotherapy by decitabine. Therefore, the epi-drugs are used as re-sensitizers of the drugs to which resistance exists. Epi-drugs in combination with targeted therapy (ER targeted therapy) have gone through clinical trials. Another phase II study was conducted with exemestane, alone and in conjunction with entinostat. The combination treatment has been found to impart progression-free survival and overall survival was also improved. Another study comprised vorinostat and tamoxifen. In yet another study with metastatic breast cancer patients, vorinostat plus bevacizumab and paclitaxel resulted in partial and complete remission.

Inhibitors of DNMT and HDACs, though have potential in breast cancer therapy, are not free from disadvantages. Their non-chromatin effects and genome wide function are the main problems (Yang et al. 2021).

Altered epigenetic processes include the enzymes DNA methyl transferases (DNMTs), histone methyl transferases (HMTs), histone demethylases (HDMs), histone deacetylases (HDACs). Of these, clinical research has been undertaken with inhibitors of DNMTs and HDACs. Abexinostat, an HDAC inhibitor has been tested in breast cancer, which may enhance drug sensitivity. Tumor consists of cells of different phenotypes, causing heterogeneity. Due to high degree of cellular differentiation and heterogeneity between the tumors, use of epi-drugs is limited in case of solid tumors. Owing to higher degree of instability and heterogeneity of tumor cells, a good knowledge of biomarkers is of much relevance that are targeted by epi-drugs.

Use of epi-drugs along with chemotherapy, immunotherapy, or other treatments face a number of challenges. A number of natural constituents act as epi-drugs; they are small-molecule inhibitors as well. Histone methylation is intricately related to carcinogenesis, therefore, research and development of small molecules which are inhibitors of histone methyl transferase may pave a path to betterment of therapy. Histone methyl transferases are broadly of two types, arginine methyl transferases, such as protein arginine methyl transferases (PMRT) and lysine methyl transferases, such as enhancer of zeste homolog 2 (EZH2). GSK3368715 is a PMRT inhibitor, showing good inhibitory action in triple-negative breast cancer. This molecule activates the p53 pathway, thereby regulating the cell cycle and triggering apoptosis.

4.4 Natural Remediation in Breast Cancer

Cancer cells constantly change to adopt themselves to therapy, hindering therapeutic efficacy. Normal or non-cancerous cells on the other hand are stable. Basically, treatment for cancer comprises surgery—to get rid of the offending growth, chemotherapy—to kill the cancer cells, and radiation therapy—to burn the cancer cells. All these treatments are symptomatic treatments that tackle the cancer cells, though normal or non-cancerous cells also are damaged. Therefore, a mode of treatment that will preferentially kill the cancer cells, sparing the normal cells, will be highly warranted. Use of food and natural supplements is a non-invasive modality, having no harmful effects. This is true for breast cancer cells too. For a promising outcome of the treatment regimen, the best strategy is to destroy the cancer cells, leaving the normal cells unaffected from the toxic effect of chemotherapy. Development of newer drugs is warranted which are nongenotoxic, but can kill cancer cells efficiently (Liu et al. 2015). Chemotherapeutic drugs generally target different genes which are aberrantly expressed in breast cancer. These drugs ultimately target adhesion molecules, cytokines, prostaglandins, and a number of enzymes. Often resistance to chemotherapeutic drugs develops, resulting in sustained survival, aberrant proliferation, and growth of cancer cells, finally resulting in invasion, angiogenesis, and metastasis. Plant-derived molecules, i.e., phytochemicals can modulate these steps, overcome chemoresistance, and increase the efficacy of these drugs, so that they work better. This has been explained in Fig. 4.2.

The most effective way to control breast cancer is prevention. This can be achieved by usage of natural products from plant origins. Phytochemicals are the active plant chemicals, which are nutrients and are found in fruits, vegetables, leaves, seeds, grains, etc. These active plant products have medicinal properties, which may provide health benefits, besides imparting nutritional values. Often these compounds reduce the risk of major chronic diseases (Liu 2004). Phytochemicals have enormous antioxidant potential, rendering protection against hazardous chemicals, radiation, etc. The phytochemicals that provide promising health benefits are polyphenols, flavonoids, isoflavonoids, anthocyanidins, phytoestrogens, terpenoids, carotenoids, limonoids, phytosterols, glucosinolates, fibers, etc. (Thakur et al. 2020). These compounds of plant origin, owing to their antioxidant, antiestrogen, and chemopreventive properties may prevent breast cancer. These active molecules may prevent formation of carcinogens, inactivate the potency of carcinogens toward the target, and thereby prevent carcinogenesis. Phytonutrients also aid in repair of the damage caused due to carcinogens. As discussed previously, estrogen supports carcinogenesis. Some of these phytochemicals help to fight breast cancer by conversion of estrogen, a cancer promoting hormone to a more protective hormone. Phytochemicals may help to fight breast cancer in other ways, some phytochemicals target signalling pathways. Phytochemicals inhibit different stages of carcinogenesis, starting from hyperplasia, dysplasia to invasive carcinoma. These have been pictorially represented in Fig. 4.3.

If the tumor microenvironment is of low pH, i.e., acidic, it supports the growth of cancer cells and helps them to become more invasive and drives metastasis. There is

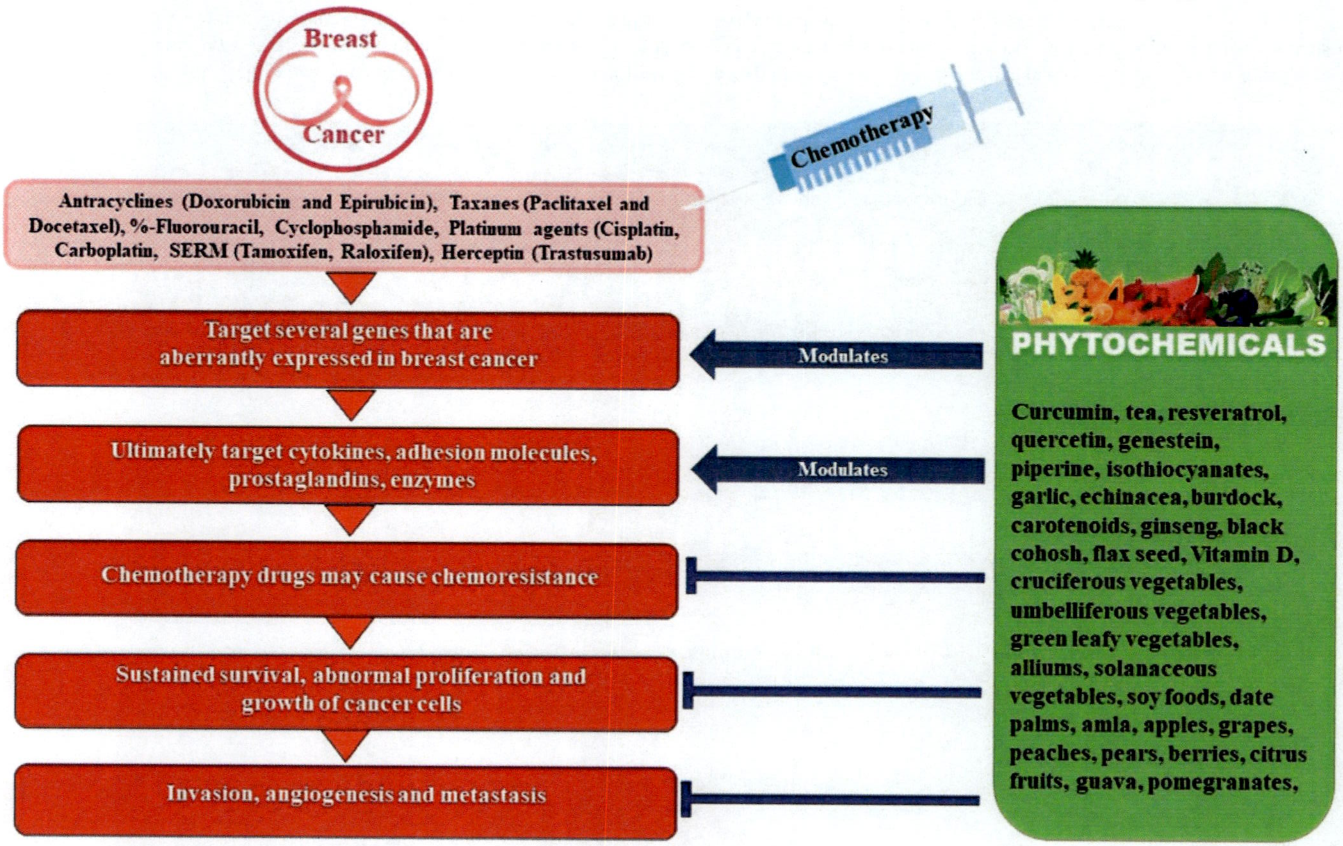

Fig. 4.2 A number of pathways are altered in breast cancer, which are regulated by several genes. These are targets for effective therapeutic regimens. Besides targeting the aberrantly expressed genes, these drugs also target cytokines and other molecules, causing chemoresistance, abnormal proliferation, and finally metastasis. Plant-derived molecules or the phytochemicals may modulate the processes

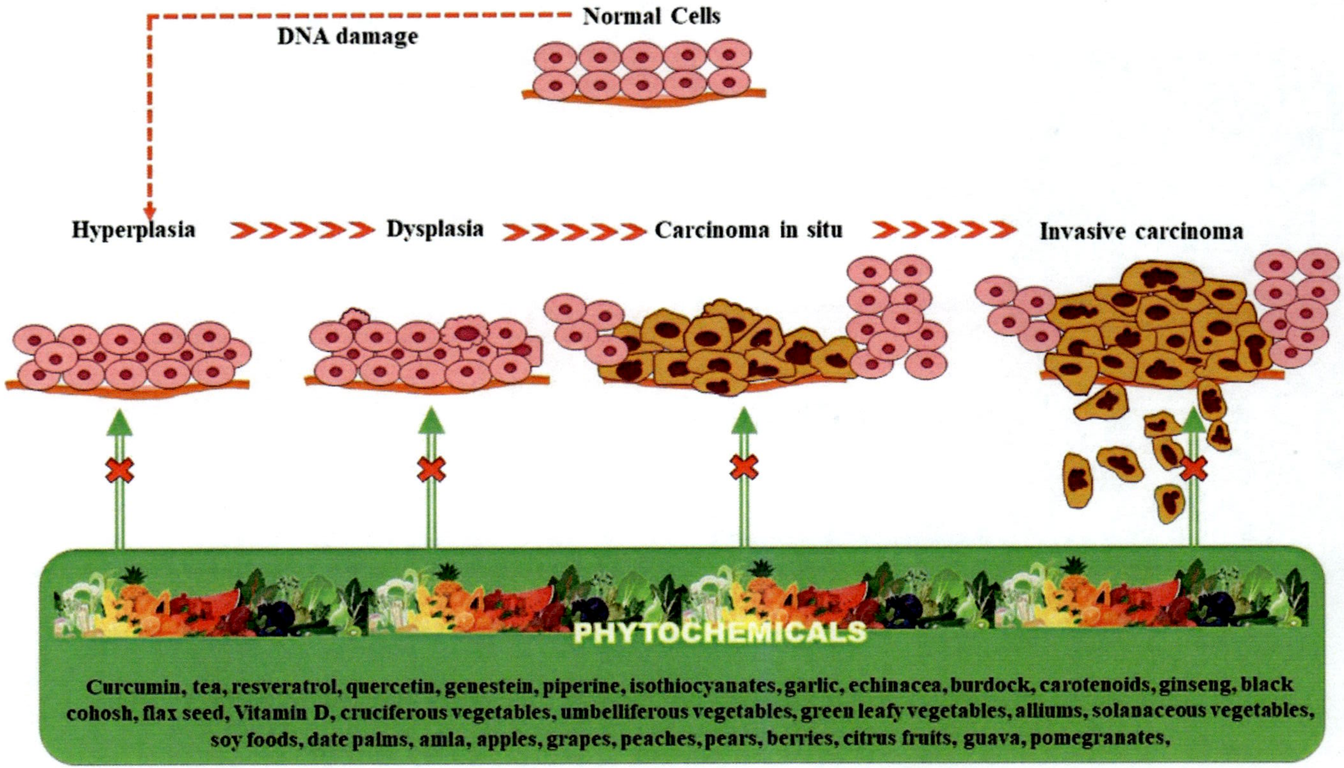

Fig. 4.3 Normal or healthy cell, when encounters damage, may be converted to a hyperplastic cell, which further may be converted to a dysplastic cell. Hyperplasia or dysplasia may not culminate into cancer. Often, they may progress to carcinoma in situ and finally invasive carcinoma. Phytochemicals may block all these stages by altering different proteins that drive signalling pathways. Phytochemicals may thus aid in cancer control

an interplay between cellular carbon metabolism and removal of acid, which is conciliated by carbonic anhydrases and some transport proteins, interstitial fluid buffering, and abnormal tumor-associated vessels. Of this carbonic anhydrase IX (CAIX) is a surface enzyme which is associated with cancer (Lee and Griffiths 2020).

Cancer may be prevented by using a combination of food and natural supplements. Curing cancer by natural means is non-invasive, i.e., they have minimal toxicity and side effects; most importantly they can be taken with any other medicine. Breast cancer, like all other cancers, may be cured by natural means, (a) by changing the acidic environment of the body to alkaline environment, as proliferation of cancer cells is checked in an alkaline environment, (b) by boosting up immunity by consumption of several natural products, these immune boosters aid in killing cancer cells. There are a number of phytonutrients which can prevent cancer, as discussed below. A detailed discussion of the beneficial effects of plant-derived molecules in cancer prevention and treatment can be found in (Roy and Datta 2019). We discuss here some of these phytochemicals in relation to breast cancer.

1. Alliums: Allium vegetables contain organosulfur compounds, which aid in prevention of cancer. Onions contain the flavonoids kaempferol, myricetin, and quercetin having strong antioxidant properties (Pourzand et al. 2016). Consumption of certain allium vegetables, in particular leek and garlic, may reduce the risk of breast cancer in high-risk population (Roy et al. 2015).
2. Allspice: Allspice (*Pimenta dioica*) is also known as Jamaican pepper or pimento. The active part of this plant product can enhance the efficacy of chemotherapeutic drugs. Allspice reduces the viability and growth of breast cancer cells. This compound inhibits Akt/mTOR and increased cytotoxicity when used in conjunction with a chemotherapy drug rapamycin. Studies show that in breast cancer cells, antitumor and chemopreventive activity of allspice is due to mTOR inhibition (Zhang et al. 2015).
3. Amla: *Phyllanthus emblica* (Indian Gooseberry), popularly known as Amla has cancer preventive and antitumor properties. This fruit is rich in polyphenols and hydrolysable tannins, which are potential antioxidants. These along with other tannins prevent mutagenesis and lipid peroxidation caused due to carcinogens and reactive oxygen species. Gallotannin elicits significant response against triple-negative breast tumors (Zhao et al. 2015).
4. Apples, grapes, peaches, pears: Apples belong to the Rosaceae family, fruits in which are rich in phytochemicals, that have potential to fight against breast cancer. These phytochemicals possess antioxidant properties and inhibit cell proliferation. Pears are also effective in breast cancer (Farvid et al. 2019). Two serving of peaches every week reduces the risk of estrogen receptor-negative breast cancer by 41% (Farvid et al. 2019).
5. Berries: Berries contain anthocyanins, flavonoids, and phenolic acids, possessing anticancer properties. They are potent inducer of programmed cell death, inhibit proliferation of cells and are modulators of cell signalling.

Anticarcinogenic properties in berries are also noteworthy. Blueberries, blackberries, raspberries, and strawberries are known for their anti-inflammatory and antioxidant attributes, that could aid in breast cancer prevention. Growth of breast cancer cells can be efficiently checked by blueberries and blackberries (Kristo et al. 2016).

6. Burdock (*Arctium lappa*): Active ingredient of burdock is arctigenin. The phenolic compound tannin, which stimulates microphage action and is responsible for immunomodulatory properties, is present in burdock. This product of plant origin is used in treating arthritis, tonsillitis, and measles. The antitumor potential has been discovered later, particularly for breast tumor, ovary, bladder, malignant melanoma, lymphoma, and pancreatic cells. Active compounds in this phytochemical can alter the oncogenes (Shareef et al. 2016). The root is flavonoid rich and restricts propagation of cancer.
7. Black cohosh (*Cimicifuga racemose*): This plant is of North American origin, where it is used to treat menopausal signs, pre-menstrual discomfort, dysmenorrhea, etc. Triterpene glycoside is a major component, however, resins and caffeic, isoferulic and fukinolic are also present, though in trace amount. A synergistic effect of chemotherapeutic drugs and black cohosh has been reported (Rockwell et al. 2005).
8. Black pepper: Piperine is a potent inhibitor of proliferation and an inducer of apoptosis, It does so by activating caspase-3 and PARP cleavage. Piperine also suppresses EGF-induced MMP-9 expression by interfering with ERK1/2, p38 and MAPK. Cell migration is significantly reduced. The sensitization of human epidermal growth factor receptor (HER)2 is often overexpressed in breast cancer cells, and sensitization of these cells to paclitaxel was augmented by piperine (Do et al. 2013). In addition, among all the compounds derived from plants that were studied, piperine is a potent adjuvant that enhances the efficacy of TNF-related and apoptosis-inducing therapies in triple-negative breast cancer cells. This works via inhibition of survivin and p65 phosphorylation (Abdelhamed et al. 2014). In triple-negative breast cancer, piperine induces apoptosis through mitochondrial pathway.
9. Basil: Flavonoids protect against cancer. Carnosol, an active ingredient of basil, inhibits carcinogenesis of hormone receptor-positive (ER+/PR+) breast cancer cells. Ocimum genus or basil inhibits tumor growth by decreasing basement membrane disintegration, angiogenesis, and MMP-2 and MMP-9 activities (Nangia-Makker et al. 2013).
10. Capsicum: Capsaicin is the active molecules in Capsicum. Cell death due to these molecules is due to up-regulation of c-Fos and RIP3 (Wu et al. 2014). Suppression of cell proliferation of breast cancer cells is facilitated by the extract. Expression of the epithelial marker E-cadherin is increased with a concomitant decrease of MMP-2 and MMP-9, and the secretion of MMP-2 and MMP-9 was decreased by the extract (Kim et al. 2014).
11. Cinnamon: *Cinnamomum cassia* (L.) J. Presl (cinnamon), main ingredient from Cinnamon, displayed a promising role on the prevention and treatment of breast cancer. Anticancer effect of cinnamaldehyde was proved in in vitro studies.

Modulation of PI3K-Akt pathway was responsible for this effect, leading to induction of apoptosis. Inhibition of cell proliferation, cell migration, and invasion are also the main attributes of cinnamon (Liu et al. 2020a).

12. Citrus fruits: Phytochemicals in citrus fruits are coumarins, limonoids, limonene, and Vitamin C. Of these, Vitamin C is an excellent antioxidant that protects cell membrane and DNA from oxidative stress. It is also capable of scavenging and reducing nitrite, thereby preventing formation of nitrosamines, which is a potent carcinogen. Consumption of citrus fruits enhances the activity of detoxifying enzymes like glutathione transferase.
13. Cumin: Active ingredient of black cumin is thymoquinone, which is an anti-proliferative and pro-apoptotic compound. This active compound elicits its action through induction of p38 phosphorylation via ROS production (Woo et al. 2013). It also inhibits Akt kinases which are activated in tumor cells (Rajput et al. 2013). It inhibits the growth and proliferation of triple-negative breast cancer. Cytotoxicity of cisplatin and docetaxel has been reported to be diminished by thymoquinone (Sutton et al. 2014).
14. Cayenne: The active compound extracted from chili peppers showed potent anticarcinogenic activities in vitro and in vivo. This has been studied in breast cancer cells MDA MB 231. This active compound downregulated the expression levels of phosphorylated (p) PI3K, p Akt, Wnt and β catenin and cell viability of breast cancer cells got downregulated. This may be due to G2/M cell cycle arrest and reduced CDK8 expression (Wu et al. 2020).
15. Carotenoids: Consumption of carotenoids may inhibit tumor progression (Eliassen et al. 2015). These are good antioxidants and protect cells from oxidative damage. They boost the immune system. Carotenoids may reduce the risk of estrogen receptor-negative (ER-) breast cancers.
16. Cruciferous vegetables: Consumption of cruciferous vegetables may reduce the risk of breast cancer (Liu and Lv 2013; Lin et al. 2017), in particular consumption of broccoli and cauliflower is beneficial. An inverse association of consumption of cruciferous vegetable with risk of breast cancer has been seen in premenopausal women.
17. Date palms: Date extracts showed cytotoxic effects on MDA-MB-231, which is a triple-negative breast cancer cell line. These extracts exhibit their effects through EGFR/ERK/FAK pathway, SRC family of kinases, and calcium signalling. Therefore, date extracts could be a therapeutic agent for breast cancer (Khattak et al. 2020).
18. Echinacea: Echinacea, belongs to a family Asteraceae. This is rich in flavonoids, which is an immune stimulant. Flavonoids enhance the phagocytosis by macrophages and the action of natural killer cells. They diminish the harmful consequences of radiotherapy and chemotherapy.
19. Flax seed: It is a rich source of dietary fibers or lignans. It contains phytoestrogens. Flax seeds and secoisolariciresinol diglucoside reduce malignancies, and anticancer activities have been reported (Calado et al. 2018).
20. Grapes: Grapes contain catechin, quercetin, resveratrol in abundance, which are all polyphenols. Besides, they contain fisetin and lupeol, having anticancer

properties. Seeds of the grapes and red grapes are rich in resveratrol, which are aromatase inhibitors, that inhibit growth-stimulatory actions. Resveratrol works well on estrogen-dependent breast cancer. This molecule modulates signal transduction pathways which control cell division, inflammation, apoptosis, angiogenesis, and metastasis. Therefore, resveratrol works on all stages of development of cancer, initiation, promotion, and progression (Sinha et al. 2016).

21. Guava: Carotenoids, copper, fibers, flavonoids, folic acid, lycopene, manganese, potassium, Vitamins A, and Vitamin C are found in abundance in guava. Many of these, owing to their antioxidant properties, counter breast cancer (Liu et al. 2020b), by eliminating free radicals.
22. Ginger: Active ingredients of ginger oleoresin (*Zingiber officinale* Roscoe) elicited anti-tumor activity; besides it has anti-inflammatory, antioxidant, and antimicrobial properties. Gingerol induces apoptosis in in vivo models and in triple-negative breast cancer cells in vitro. Gingerol activates caspase-3 and inhibits metastasis (Martin et al. 2017). Ginger has been found to enhance adiponectin, NO and GPx and reduce MDA in obese women diagnosed with breast cancer (Karimi and Roshan 2013).
23. Garlic: Garlic (*Allium sativum*) is anticancer, antimicrobial, antiarthritic. Antithrombotic, hypolipidaemic, anti-inflammatory, and hypoglycemic; this product is an immune booster. Anti-breast cancer effects are also attributed to ajoene, apigenin, enterolactone, allicin, diallyl disulfide and diallyl trisulfide, diallyl tetrasulfide, and dipropyltetrasulfide. Of these compounds, ajoene interferes with protein folding in triple-negative (ER−/PR−/HER2−) breast cancer cells, leading to cell death. Diallyl disulfide is an apoptotic inducer in ER+/PR+, as well as triple-negative breast cancer cells. Diallyl trisulfide prevents cell invasion and metastasis, suppressing the progression of breast cancer (Roy et al. 2015).
24. Green leafy vegetables: Fruit and vegetable consumption improves the overall survival among breast cancer patients (Roy et al. 2015). Non-starchy vegetables may diminish the risk of estrogen receptor-negative (ER−) breast cancer. Kale, a dark green leafy vegetable contains abundant antioxidants. Therefore, these vegetables boost the antioxidant defenses, preventing the free radicals from damaging DNA, thus preventing cancer. Vitamin C in kale is also a powerful antioxidant and helps to inhibit formation of carcinogens.
25. Ginseng: (*Panax ginseng*): This compound restricts cancer risk, blocks the propagation and metastasis of cancer cells. Beneficial effects of active compounds of panax ginseng include restoration of natural killer cells, which are impaired during chemotherapy and radiotherapy. These compounds also induce macrophages and enhance formation of antibodies (Hong et al. 2021).
26. Green tea: Epigallocatechin (EGGG) is known for its anticancer and antimutagenic activity. Green tea polyphenols check the division of cancer cells. Besides, they are apoptotic inducer (Zaveri 2006). These compounds are immune booster, inhibit metastasis and angiogenesis. Green tea contains catechins that have benefits in breast cancer prevention. A study (Xiang et al.

2016) suggests that green tea may be quite effective in slowing tumor growth and modulating other biological processes that contribute to breast cancer (Xiang et al. 2016). Another study suggests that consumption of five cups of green or black tea per week may be associated with decrease in breast cancer risk up to 15% (Zhang et al. 2020).

27. Lavender: The essential oils in Lavender are due to *Lavandula angustifolia* (common lavender), *Lavandula latifolia* (spike lavender), and *Lavandula intermedia*, a sterile hybrid of *Lavandula angustifolia* and *Lavandula latifolia* (lavandin). Complications in cancer can be tackled by lavender therapy. Inhalation of aromatherapy using lavender oil for breast cancer patients may aid in decreasing anxiety levels before surgery (Beyliklioğlu and Arslan 2019).
28. Mint: It contains perillyl alcohol, which inhibits the growth of cancer cells. Peppermint oil may protect against breast cancer by inducing apoptosis and arresting cell cycle. However human trials are needed (Rouhi 2014).
29. Onion: Cancer preventive potential has also been shown by onion (*Allium cepa*) and scallion (*Allium fistulosum*). They contain sulfur containing compounds, and preventive effects are due to these compounds (Nicastro et al. 2015). Onion also contains quercetin and fisetin which contribute to the chemopreventive properties.
30. Oregano: Oregano (*Origanum vulgare*), a herb belonging to the mint family has antioxidant, anti-inflammatory, antibacterial, and antifungal properties. It contains volatile oils that include thymol, eugenol, carvacrol, and limonene. In vitro studies indicate a tumor-suppressive effect of oregano in breast cancer in an animal model. Mexican oregano is even better than oregano, as it contains more chemopreventive compounds. Luteolin present in Mexican oregano suppresses the proliferation of triple-negative breast cancer cells and metastasis and reduces ER+/PR+ cell viability in a dose- and time-dependent manner. Luteolin also inhibits angiogenesis and aromatase activity (Kubatka et al. 2017).
31. Parsley: Parsley contains a flavonoid apigenine, which is chemopreventive. It decreases tumor size in some aggressive form of breast cancer. Apigenin may be a promising non-toxic cancer treatment (Danciu et al. 2021).
32. Pomegranates: Chemopreventive attributes of pomegranates are due to the presence of anthocyanins, ellagic acid, luteolin, punicic acid, ursolic acid, etc. Presence of tannins and flavonoids helps to combat breast cancer (Sharma et al. 2017).
33. Rosemary: It contains carnosol, which slows down cancer cell growth, boosts immune function, and limits the production of cancer related hormones. Rosemary has antioxidant properties. Several studies suggest that rosemary extract may inhibit tumor growth by inhibiting cancerous cells from replicating. One study found that rosemary, on its own and in combination with curcumin, helped prevent breast cancer. Anticancer potential is mainly due to carnosic acid and rosmarinic acid, and their effects on key signalling molecules. It was observed that extracts show antitumor activity against breast cancer cells due to the downregulation of ER-α and HER2 receptors. besides these, rosemary extracts

potentiated the effect of chemotherapy drugs like tamoxifen, trastuzumab, and paclitaxel (González-Vallinas et al. 2014).

34. Soy foods: Epidemiological studies conducted in USA and China have shown that soy intake significantly reduces recurrence of breast cancer and increases survival (Messina 2016). Literature shows that consumption of soy foods appreciably reduces risk of breast cancer—even in patients with estrogen receptor-positive cancer.
35. Solanaceous vegetables: These vegetables contain lycopene, capsaicin, and anthocyanin. These active compounds inhibit cell proliferation. Capsaicin found in capsicum also belongs to this group of vegetables. Capsaicin is known for its anticarcinogenic, antimutagenic, and chemopreventive attributes (Roy et al. 2015).
36. Saffron: Major compounds of Saffron are crocin, crocetin, picrocrocin, and safranal. It has been reported that saffron extract decreased cell viability and induced apoptosis through activation of caspases and increased Bax expression (Lambrianidou et al. 2020). VEGF-A and VEGFR-2, two important biomarkers of angiogenesis get inhibited by saffron in breast cancer cells.
37. Thyme: It contains terpenoids, which is a good antioxidant and protects from cancer. The aromatic compounds and oily components of thyme have shown cytotoxic activity against breast cancer (Sertel et al. 2011).
38. Turmeric: (Curcuma longa): Active ingredient of turmeric is Curcumin, which is found in the rhizome of the plant, also in the root stock. It is known for its anticancer properties, which is mainly due to the phenolic compounds. Curcumin is also known for its antioxidant, anti-inflammatory, analgesic, and antiseptic action from ancient times. It has inhibitory action on all the stages in carcinogenesis, initiation, promotion and progression may be inhibited by curcumin. Production of carcinogens like nitrosamine may be inhibited by curcumin. Antioxidant action of the body is enhanced by curcumin. It has anticancer effects on many types of cancer, including breast cancer. Curcumin lowers HIF-1α and HIF-2α, inhibits cell invasion and 12-O-tetradecanoylphorbol-13-acetate (TPA)-induced matrix metalloproteinase MMP-9 expression. The mechanism of action is most probably through the suppression of protein kinase C-α (PKC-α), mitogen-activated protein kinase (MAPK), and nuclear factor-κB/active protein-1 (NF-κB/AP-1) pathway. Suppressed level of the epithelial marker E-cadherin gets restored by curcumin. Most aggressive form of breast cancer is triple-negative breast cancer. Curcumin works by inhibition of EGFR signalling pathway (Cine et al. 2013). Triple-negative breast cancer which is resistant to retinoic acid gets sensitized by curcumin (Mukherjee et al. 2014).
39. Umbelliferous vegetables: Umbelliferous vegetables like celery, parsnip, carrot, etc. possess anticancer properties. Phytochemicals in these vegetables have the ability to inhibit aromatase activity, which is responsible for diminishing the growth-stimulatory effects in estrogen-dependent breast cancer (Roy et al. 2015).

40. Vitamin D: Anticancer activity of Vitamin D has been documented, and vitamin D deficiency may cause higher incidence of breast cancer (Atoum and Alzoughool 2017). Vitamin D, in conjunction with conventional cancer therapies could have a promising role in breast cancer therapy, though more investigation is required to establish the role of this molecule to slow down the growth of tumor and progression of the disease (Vanhevel et al. 2022).
41. Whole grains: Whole grains include wheat, brown rice, barley, quinoa, and rye. These contain essential nutrients like fiber, vitamins, minerals, etc. These are good antioxidants and have cancer fighting abilities. Regular and adequate consumption of whole grains appreciably lowers the risk of breast cancer development (Xiao et al. 2018). Whole grains are also a good source of some essential non-nutrients, such as phytoestrogens, phenolic acids, and lignans. These compounds show their protective roles against cancer due to their antioxidant properties. They inhibit cell proliferation and angiogenesis; modulate hormonal pathways and are good inducers of apoptosis (Mourouti et al. 2016).

4.5 Conclusion

Breast cancer is the most common cancer among women. Risk of cancer is a complex one, and there are many risk factors. The interesting fact is that the diet we consume can have a great impact on the cancer risk. This is true for all cancers, particularly for breast cancer. Awareness can reduce the suffering due to cancer by educating patients and supporting cancer prevention. Therefore, accurate and understandable information should be disseminated among all classes of people. Breast cancer awareness is an effort to reduce the stigma of breast cancer through education on symptoms and treatments. Broader knowledge will lead to earlier detection of breast cancer, which is associated with higher long-term survival rates. Breast cancer advocacy and awareness efforts are important. Educational campaigns are needed for increasing awareness of breast cancer. Breast cancer culture or pink ribbon culture is a part of larger women's health movement. More than 30% of cancer cases could be prevented by modifying lifestyle or avoiding the key risk factors. "Prevent the preventable" is the way to control cancer. Many cancers could be prevented by changing lifestyle, like eating more vegetables, fruits and whole grains, eating less amount of refined carbohydrates and meat, maintaining a healthy weight, exercising, minimizing sunlight exposure, and being vaccinated against some infectious diseases. Therefore, a thorough knowledge about the goodness of plant-derived products is highly warranted.

References

Abdelhamed S, Yokoyama S, Refaat A, Ogura K, Yagita H, Awale S, Saiki I (2014) Piperine enhances the efficacy of TRAIL-based therapy for triple-negative breast cancer cells. Anticancer Res 34(4):1893–1899

Atoum M, Alzoughool F (2017) Vitamin D and breast cancer: latest evidence and future steps. Breast Cancer (Auckl) 11:1178223417749816. https://doi.org/10.1177/1178223417749816

Barneh F, Mirzaie M, Nickchi P et al (2019) Integrated use of bioinformatic resources reveals that co-targeting of histone deacetylases, IKBK and SRC inhibits epithelial-mesenchymal transition in cancer. Brief Bioinform 20(2):717–731. https://doi.org/10.1093/bib/bby030

Baylin SB, Jones PA (2016) Epigenetic determinants of cancer. Cold Spring Harb Perspect Biol 8(9):a019505. https://doi.org/10.1101/cshperspect.a019505

Beylikli̇oğlu A, Arslan S (2019) Effect of lavender oil on the anxiety of patients before breast surgery. J Perianesth Nurs 34(3):587–593. https://doi.org/10.1016/j.jopan.2018.10.002. Epub 2019 Jan 16

Bian X, Liang Z, Feng A, Salgado E, Shim H (2018) HDAC inhibitor suppresses proliferation and invasion of breast cancer cells through regulation of miR-200c targeting CRKL. Biochem Pharmacol 147:30–37. https://doi.org/10.1016/j.bcp.2017.11.008. Epub 2017 Dec 1

Blatt J, Corey SJ (2013) Drug repurposing in pediatrics and pediatric hematology oncology. Drug Discov Today 18(1–2):4–10. https://doi.org/10.1016/j.drudis.2012.07.009

Borges S, Döppler H, Perez EA, Andorfer CA, Sun Z, Anastasiadis PZ, Thompson E, Geiger XJ, Storz P (2013) Pharmacologic reversion of epigenetic silencing of the PRKD1 promoter blocks breast tumor cell invasion and metastasis. Breast Cancer Res 15(2):R66. https://doi.org/10.1186/bcr3460

Buocikova V, Rios-Mondragon I, Pilalis E, Chatziioannou A, Miklikova S, Mego M, Pajuste K, Rucins M, Yamani NE, Longhin EM, Sobolev A, Freixanet M, Puntes V, Plotniece A, Dusinska M, Cimpan MR, Gabelova A, Smolkova B (2020) Epigenetics in breast cancer therapy-new strategies and future nanomedicine perspectives. Cancers (Basel) 12(12):3622. https://doi.org/10.3390/cancers12123622

Calado A, Neves PM, Santos T, Ravasco P (2018) The effect of flaxseed in breast cancer: a literature review. Front Nutr 5:4. https://doi.org/10.3389/fnut.2018.00004

Chen J, Tian W, Cai H, He H, Deng Y (2012) Down-regulation of microRNA-200c is associated with drug resistance in human breast cancer. Med Oncol 29(4):2527–2534. https://doi.org/10.1007/s12032-011-0117-4. Epub 2011 Nov 19

Chew HK (2001) Adjuvant therapy for breast cancer: who should get what? West J Med 174(4): 284–287. https://doi.org/10.1136/ewjm.174.4.284

Cine N, Limtrakul P, Sunnetci D, Nagy B, Savli H (2013) Effects of curcumin on global gene expression profiles in the highly invasive human breast carcinoma cell line MDA-MB 231: a gene network-based microarray analysis. Exp Ther Med 5(1):23–27. https://doi.org/10.3892/etm.2012.754. Epub 2012 Oct 19

Danciu C, Cioanca O, Watz Farcaş C, Hancianu M, Racoviceanu R, Muntean D, Zupko I, Oprean C, Tatu C, Paunescu V, Proks M, Diaconeasa Z, Soica C, Pinzaru I, Dehelean C (2021) Botanical therapeutics (part II): antimicrobial and in vitro anticancer activity against MCF7 human breast cancer cells of chamomile, parsley and celery alcoholic extracts. Anti Cancer Agents Med Chem 21(2):187–200. https://doi.org/10.2174/1871520620666200807213734

De Mattos-Arruda L, Bottai G, Nuciforo PG, Di Tommaso L, Giovannetti E, Peg V et al (2015) MicroRNA-21 links epithelial-to-mesenchymal transition and inflammatory signals to confer resistance to neoadjuvant trastuzumab and chemotherapy in HER2-positive breast cancer patients. Oncotarget 6(35):37269–37280. https://doi.org/10.18632/oncotarget.5495

de Oca RM, Gurard-Levin ZA, Berger F, Rehman H, Martel E, Corpet A, de Koning L, Vassias I, Wilson LO, Meseure D, Reyal F, Savignoni A, Asselain B, Sastre-Garau X, Almouzni G (2015) The histone chaperone HJURP is a new independent prognostic marker for luminal A breast carcinoma. Mol Oncol 9(3):657–674. https://doi.org/10.1016/j.molonc.2014.11.002. Epub 2014 Nov 20

Do MT, Kim HG, Choi JH, Khanal T, Park BH, Tran TP, Jeong TC, Jeong HG (2013) Antitumor efficacy of piperine in the treatment of human HER2-overexpressing breast cancer cells. Food Chem 141(3):2591–2599. https://doi.org/10.1016/j.foodchem.2013.04.125. Epub 2013 May 24

Eliassen AH, Liao X, Rosner B, Tamimi RM, Tworoger SS, Hankinson SE (2015) Plasma carotenoids and risk of breast cancer over 20 years of follow-up. Am J Clin Nutr 101(6): 1197–1205. https://doi.org/10.3945/ajcn.114.105080. Epub 2015 Apr 15

Falahi F, van Kruchten M, Martinet N et al (2014) Current and upcoming approaches to exploit the reversibility of epigenetic mutations in breast cancer. Breast Cancer Res 16:412. https://doi.org/10.1186/s13058-014-0412-z

Farvid MS, Chen WY, Rosner BA, Tamimi RM, Willett WC, Eliassen AH (2019) Fruit and vegetable consumption and breast cancer incidence: repeated measures over 30 years of follow-up. Int J Cancer 144(7):1496–1510. https://doi.org/10.1002/ijc.31653. Epub 2018 Dec 19

Ganesan A, Arimondo PB, Rots MG et al (2019) The timeline of epigenetic drug discovery: from reality to dreams. Clin Epigenetics 11:174. https://doi.org/10.1186/s13148-019-0776-0

González-Vallinas M, Molina S, Vicente G, Sánchez-Martínez R, Vargas T, García-Risco MR, Fornari T, Reglero G, Ramírez de Molina A (2014) Modulation of estrogen and epidermal growth factor receptors by rosemary extract in breast cancer cells. Electrophoresis 35(11): 1719–1727. https://doi.org/10.1002/elps.201400011. Epub 2014 Mar 20

Handy DE, Castro R, Loscalzo J (2011) Epigenetic modifications: basic mechanisms and role in cardiovascular disease. Circulation 123(19):2145–2156. https://doi.org/10.1161/CIRCULATIONAHA.110.956839

Hare SH, Harvey AJ (2017) mTOR function and therapeutic targeting in breast cancer. Am J Cancer Res 7(3):383–404

Hong H, Baatar D, Hwang SG (2021) Anticancer activities of ginsenosides, the main active components of ginseng. Evid Based Complement Alternat Med 2021:8858006. https://doi.org/10.1155/2021/8858006

Hyun K, Jeon J, Park K, Kim J (2017) Writing, erasing and reading histone lysine methylations. Exp Mol Med 49(4):e324. https://doi.org/10.1038/emm.2017.11

Karimi N, Roshan VD (2013) Change in adiponectin and oxidative stress after modifiable lifestyle interventions in breast cancer cases. Asian Pac J Cancer Prev 14(5):2845–2850. https://doi.org/10.7314/apjcp.2013.14.5.2845

Khattak MNK, Shanableh A, Hussain MI, Khan AA, Abdulwahab M, Radeef W, Samreen MH (2020) Anticancer activities of selected Emirati Date (*Phoenix dactylifera* L.) varieties pits in human triple negative breast cancer MDA-MB-231 cells. Saudi J Biol Sci 27(12):3390–3396. https://doi.org/10.1016/j.sjbs.2020.09.001. Epub 2020 Sep 6

Kim HA, Kim MS, Kim SH, Kim YK (2014) Pepper seed extract suppresses invasion and migration of human breast cancer cells. Nutr Cancer 66(1):159–165. https://doi.org/10.1080/01635581.2014.853814. Epub 2013 Dec 16

Kristo AS, Klimis-Zacas D, Sikalidis AK (2016) Protective role of dietary berries in cancer. Antioxidants (Basel) 5(4):37. https://doi.org/10.3390/antiox5040037

Kubatka P, Kello M, Kajo K, Kruzliak P, Výbohová D, Mojžiš J, Adamkov M, Fialová S, Veizerová L, Zulli A, Péč M, Statelová D, Grančai D, Büsselberg D (2017) Oregano demonstrates distinct tumour-suppressive effects in the breast carcinoma model. Eur J Nutr 56(3):1303–1316. https://doi.org/10.1007/s00394-016-1181-5. Epub 2016 Feb 23

Lambrianidou A, Koutsougianni F, Papapostolou I, Dimas K (2020) Recent advances on the anticancer properties of saffron (*Crocus sativus* L.) and its major constituents. Molecules 26(1):86. https://doi.org/10.3390/molecules26010086

Lee SH, Griffiths JR (2020) How and why are cancers acidic? Carbonic anhydrase IX and the homeostatic control of tumour extracellular pH. Cancers (Basel) 12(6):1616. https://doi.org/10.3390/cancers12061616

Li A, Liu Z, Li M, Zhou S, Xu Y, Xiao Y, Yang W (2016) HDAC5, a potential therapeutic target and prognostic biomarker, promotes proliferation, invasion and migration in human breast cancer. Oncotarget 7(25):37966–37978. https://doi.org/10.18632/oncotarget.9274. Erratum in: Oncotarget 2017 May 2;8(18):30619–30620

Li Z, Zou W, Zhang J, Zhang Y, Xu Q, Li S, Chen C (2020) Mechanisms of CDK4/6 inhibitor resistance in luminal breast cancer. Front Pharmacol 11:580251. https://doi.org/10.3389/fphar.2020.580251

Lin T, Zirpoli GR, McCann SE, Moysich KB, Ambrosone CB, Tang L (2017) Trends in cruciferous vegetable consumption and associations with breast cancer risk: a case-control study. Curr Dev Nutr 1(8):e000448. https://doi.org/10.3945/cdn.117.000448

Liu RH (2004) Potential synergy of phytochemicals in cancer prevention: mechanism of action. J Nutr 134(12 Suppl):3479S–3485S. https://doi.org/10.1093/jn/134.12.3479S

Liu X, Lv K (2013) Cruciferous vegetables intake is inversely associated with risk of breast cancer: a meta-analysis. Breast 22(3):309–313. https://doi.org/10.1016/j.breast.2012.07.013. Epub 2012 Aug 9

Liu B, Ezeogu L, Zellmer L, Yu B, Xu N, Joshua LD (2015) Protecting the normal in order to better kill the cancer. Cancer Med 4(9):1394–1403. https://doi.org/10.1002/cam4.488. Epub 2015 Jul 14

Liu Y, An T, Wan D, Yu B, Fan Y, Pei X (2020a) Targets and mechanism used by cinnamaldehyde, the main active ingredient in cinnamon, in the treatment of breast cancer. Front Pharmacol 11: 582719. https://doi.org/10.3389/fphar.2020.582719

Liu HC, Chiang CC, Lin CH, Chen CS, Wei CW, Lin SY, Yiang GT, Yu YL (2020b) Anti-cancer therapeutic benefit of red guava extracts as a potential therapy in combination with doxorubicin or targeted therapy for triple-negative breast cancer cells. Int J Med Sci 17(8):1015–1022. https://doi.org/10.7150/ijms.40131

Locke WJ, Clark SJ (2012) Epigenome remodelling in breast cancer: insights from an early in vitro model of carcinogenesis. Breast Cancer Res 14(6):215. https://doi.org/10.1186/bcr3237

Lu L, Ju F, Zhao H, Ma X (2015) MicroRNA-134 modulates resistance to doxorubicin in human breast cancer cells by downregulating ABCC1. Biotechnol Lett 37(12):2387–2394. https://doi.org/10.1007/s10529-015-1941-y. Epub 2015 Aug 29

Mansoori B, Mohammadi A, Davudian S, Shirjang S, Baradaran B (2017) The different mechanisms of cancer drug resistance: a brief review. Adv Pharm Bull 7(3):339–348. https://doi.org/10.15171/apb.2017.041

Martin ACBM, Fuzer AM, Becceneri AB, da Silva JA, Tomasin R, Denoyer D, Kim SH, McIntyre KA, Pearson HB, Yeo B, Nagpal A, Ling X, Selistre-de-Araújo HS, Vieira PC, Cominetti MR, Pouliot N (2017) [10]-gingerol induces apoptosis and inhibits metastatic dissemination of triple negative breast cancer in vivo. Oncotarget 8(42):72260–72271. https://doi.org/10.18632/oncotarget.20139

Messina M (2016) Impact of soy foods on the development of breast cancer and the prognosis of breast cancer patients. Forsch Komplementmed 23(2):75–80. https://doi.org/10.1159/000444735. Epub 2016 Apr 12

Mithraprabhu S, Kalff A, Chow A, Khong T, Spencer A (2014) Dysregulated class I histone deacetylases are indicators of poor prognosis in multiple myeloma. Epigenetics 9(11): 1511–1520. https://doi.org/10.4161/15592294.2014.983367

Montalvo-Casimiro M, González-Barrios R, Meraz-Rodriguez MA, Juárez-González VT, Arriaga-Canon C, Herrera LA (2020) Epidrug repurposing: discovering new faces of old acquaintances in cancer therapy. Front Oncol 10:605386. https://doi.org/10.3389/fonc.2020.605386

Moo TA, Sanford R, Dang C, Morrow M (2018) Overview of breast cancer therapy. PET Clin 13(3):339–354. https://doi.org/10.1016/j.cpet.2018.02.006

Moreira-Silva F, Camilo V, Gaspar V, Mano JF, Henrique R, Jerónimo C (2020) Repurposing old drugs into new epigenetic inhibitors: promising candidates for cancer treatment? Pharmaceutics 12(5):410. https://doi.org/10.3390/pharmaceutics12050410

Morel D, Jeffery D, Aspeslagh S, Almouzni G, Postel-Vinay S (2020) Combining epigenetic drugs with other therapies for solid tumours–past lessons and future promise. Nat Rev Clin Oncol 17(2):91–107. https://doi.org/10.1038/s41571-019-0267-4

Mourouti N, Kontogianni MD, Papavagelis C, Psaltopoulou T, Kapetanstrataki MG, Plytzanopoulou P, Vassilakou T, Malamos N, Linos A, Panagiotakos DB (2016) Whole grain consumption and breast cancer: a case-control study in women. J Am Coll Nutr 35(2):143–149. https://doi.org/10.1080/07315724.2014.963899. Epub 2015 Apr 27

Mukherjee S, Mazumdar M, Chakraborty S, Manna A, Saha S, Khan P, Bhattacharjee P, Guha D, Adhikary A, Mukhjerjee S, Das T (2014) Curcumin inhibits breast cancer stem cell migration by amplifying the E-cadherin/β-catenin negative feedback loop. Stem Cell Res Ther 5(5):116. https://doi.org/10.1186/scrt506

Nangia-Makker P, Raz T, Tait L, Shekhar MP, Li H, Balan V, Makker H, Fridman R, Maddipati K, Raz A (2013) Ocimum gratissimum retards breast cancer growth and progression and is a natural inhibitor of matrix metalloproteases. Cancer Biol Ther 14(5):417–427. https://doi.org/10.4161/cbt.23762. Epub 2013 Feb 4

Naveja JJ, Dueñas-González A, Medina-Franco JL (2016) Drug repurposing for epigenetic targets guided by computational methods. In: Epi-Informatics. Elsevier, Amsterdam, pp 327–357. https://doi.org/10.1016/B978-0-12-802808-7.00012-5

Nicastro HL, Ross SA, Milner JA (2015) Garlic and onions: their cancer prevention properties. Cancer Prev Res (Phila) 8(3):181–189. https://doi.org/10.1158/1940-6207.CAPR-14-0172. Epub 2015 Jan 13

Portela A, Esteller M (2010) Epigenetic modifications and human disease. Nat Biotechnol 28:1057–1068. https://doi.org/10.1038/nbt.1685

Pourzand A, Tajaddini A, Pirouzpanah S, Asghari-Jafarabadi M, Samadi N, Ostadrahimi AR, Sanaat Z (2016) Associations between dietary *Allium* vegetables and risk of breast cancer: a hospital-based matched case-control study. J Breast Cancer 19(3):292–300. https://doi.org/10.4048/jbc.2016.19.3.292. Epub 2016 Sep 23. Erratum in: J Breast Cancer. 2018 Jun;21(2):231

Rajput S, Kumar BN, Dey KK, Pal I, Parekh A, Mandal M (2013) Molecular targeting of Akt by thymoquinone promotes G(1) arrest through translation inhibition of cyclin D1 and induces apoptosis in breast cancer cells. Life Sci 93(21):783–790. https://doi.org/10.1016/j.lfs.2013.09.009. Epub 2013 Sep 15

Rockwell S, Liu Y, Higgins SA (2005) Alteration of the effects of cancer therapy agents on breast cancer cells by the herbal medicine black cohosh. Breast Cancer Res Treat 90(3):233–239. https://doi.org/10.1007/s10549-004-4260-x

Rodríguez-Paredes M, Esteller M (2011) Cancer epigenetics reaches mainstream oncology. Nat Med 17(3):330–339. https://doi.org/10.1038/nm.2305

Roessler J, Ammerpohl O, Gutwein J, Steinemann D, Schlegelberger B, Weyer V, Sariyar M, Geffers R, Arnold N, Schmutzler R (2015) The CpG Island methylator phenotype in breast cancer is associated with the lobular subtype. Epigenomics 7(2):187–199. https://doi.org/10.2217/epi.14.74

Rouhi SZT (2014) Proliferation inhibitory effects of peppermint oil on human breast cancer cell lines. J Cancer Sci Ther 6:10. https://doi.org/10.4172/1948-5956.S1.034

Roy M, Datta A (2019) Cancer genetics and therapeutics: focus on phytochemicals. Springer, Cham

Roy M, Mukherjee A, Mukherjee S, Biswas J (2015) Phytonutrients from fruits and vegetables in breast cancer; anticancer properties of fruits and vegetables–a scientific review. In: Kunnumakkara AB (ed) Anticancer properties of fruits and vegetables. World Scientific, Singapore, pp 75–101

Royce ME, Osman D (2015) Everolimus in the treatment of metastatic breast cancer. Breast Cancer (Auckl) 9:73–79. https://doi.org/10.4137/BCBCR.S29268

Sertel S, Eichhorn T, Plinkert PK, Efferth T (2011) Cytotoxicity of Thymus vulgaris essential oil towards human oral cavity squamous cell carcinoma. Anticancer Res 31(1):81–87

Shareef M, Ashraf MA, Sarfraz M (2016) Natural cures for breast cancer treatment. Saudi Pharm J 24(3):233–240. https://doi.org/10.1016/j.jsps.2016.04.018. Epub 2016 May 5

Sharma P, McClees SF, Afaq F (2017) Pomegranate for prevention and treatment of cancer: an update. Molecules 22(1):177. https://doi.org/10.3390/molecules22010177

Sher G, Salman NA, Khan AQ, Prabhu KS, Raza A, Kulinski M, Dermime S, Haris M, Junejo K, Uddin S (2022) Epigenetic and breast cancer therapy: promising diagnostic and therapeutic applications, seminars in cancer biology. Semin Cancer Biol 83:152–165. https://doi.org/10.1016/j.semcancer.2020.08.009

Shim JS, Liu JO (2014) Recent advances in drug repositioning for the discovery of new anticancer drugs. Int J Biol Sci 10(7):654–663. https://doi.org/10.7150/ijbs.9224

Sinha D, Sarkar N, Biswas J, Bishayee A (2016) Resveratrol for breast cancer prevention and therapy: preclinical evidence and molecular mechanisms. Semin Cancer Biol 40–41:209–232

Stefansson OA, Esteller M (2013) Epigenetic modifications in breast cancer and their role in personalized medicine. Am J Pathol 183:1052–1063. https://doi.org/10.1016/j.ajpath.2013.04.033

Sutton KM, Greenshields AL, Hoskin DW (2014) Thymoquinone, a bioactive component of black caraway seeds, causes G1 phase cell cycle arrest and apoptosis in triple-negative breast cancer cells with mutant p53. Nutr Cancer 66(3):408–418. https://doi.org/10.1080/01635581.2013.878739. Epub 2014 Feb 28

Thakur M, Singh K, Khedkar R (2020) Phytochemicals: extraction process, safety assessment, toxicological evaluations, and regulatory issues. In: Prakash B (ed) Functional and preservative properties of phytochemicals. Academic Press, Amsterdam, pp 341–361. https://doi.org/10.1016/B978-0-12-818593-3.00011-7. ISBN 9780128185933

Tung N, Domchek SM, Stadler Z, Nathanson KL, Couch F, Garber JE, Offit K, Robson ME (2016) Counselling framework for moderate-penetrance cancer-susceptibility mutations. Nat Rev Clin Oncol 13(9):581–588. https://doi.org/10.1038/nrclinonc.2016.90. Epub 2016 Jun 14

Turashvili G, Brogi E (2017) Tumor heterogeneity in breast cancer. Front Med (Lausanne) 4:227. https://doi.org/10.3389/fmed.2017.00227

Vanhevel J, Verlinden L, Doms S, Wildiers H, Verstuyf A (2022) The role of vitamin D in breast cancer risk and progression. Endocr Relat Cancer 29(2):R33–R55. https://doi.org/10.1530/ERC-21-0182

Verma M, Kumar V (2018) Chapter 21–epigenetic drugs for cancer and precision medicine. In: Moskalev A, Vaiserman AM (eds) Translational epigenetics, epigenetics of aging and longevity, vol 4. Academic Press, Cambridge, pp 439–451

Vogelstein B, Kinzler KW (2004) Cancer genes and the pathways they control. Nat Med 10(8):789–799. https://doi.org/10.1038/nm1087

Wang Z, Wang N, Liu P, Chen Q, Situ H, Xie T, Zhang J, Peng C, Lin Y, Chen J (2014) MicroRNA-25 regulates chemoresistance-associated autophagy in breast cancer cells, a process modulated by the natural autophagy inducer isoliquiritigenin. Oncotarget 5(16):7013–7026. https://doi.org/10.18632/oncotarget.2192

Woo CC, Hsu A, Kumar AP, Sethi G, Tan KH (2013) Thymoquinone inhibits tumor growth and induces apoptosis in a breast cancer xenograft mouse model: the role of p38 MAPK and ROS. PLoS One 8(10):e75356. https://doi.org/10.1371/journal.pone.0075356

Wu TT, Peters AA, Tan PT, Roberts-Thomson SJ, Monteith GR (2014) Consequences of activating the calcium-permeable ion channel TRPV1 in breast cancer cells with regulated TRPV1 expression. Cell Calcium 56(2):59–67. https://doi.org/10.1016/j.ceca.2014.04.006. Epub 2014 May 9

Wu D, Jia H, Zhang Z, Li S (2020) Capsaicin suppresses breast cancer cell viability by regulating the CDK8/PI3K/Akt/Wnt/β-catenin signaling pathway. Mol Med Rep 22(6):4868–4876. https://doi.org/10.3892/mmr.2020.11585. Epub 2020 Oct 11

Xiang LP, Wang A, Ye JH, Zheng XQ, Polito CA, Lu JL, Li QS, Liang YR (2016) Suppressive effects of tea Catechins on breast cancer. Nutrients 8(8):458. https://doi.org/10.3390/nu8080458

Xiao Y, Ke Y, Wu S, Huang S, Li S, Lv Z, Yeoh EK, Lao X, Wong S, Kim JH, Colditz GA, Tamimi RM, Su X (2018) Association between whole grain intake and breast cancer risk: a systematic review and meta-analysis of observational studies. Nutr J 17(1):87. https://doi.org/10.1186/s12937-018-0394-2

Yang T, Yang Y, Wang Y (2021) Predictive biomarkers and potential drug combinations of epi-drugs in cancer therapy. Clin Epigenetics 13:113. https://doi.org/10.1186/s13148-021-01098-2

Zaveri NT (2006) Green tea and its polyphenolic catechins: medicinal uses in cancer and noncancer applications. Life Sci 78(18):2073–2080. https://doi.org/10.1016/j.lfs.2005.12.006. Epub 2006 Jan 30

Zhang L, Shamaladevi N, Jayaprakasha GK, Patil BS, Lokeshwar BL (2015) Polyphenol-rich extract of Pimenta dioica berries (allspice) kills breast cancer cells by autophagy and delays growth of triple negative breast cancer in athymic mice. Oncotarget 6(18):16379–16395. https://doi.org/10.18632/oncotarget.3834

Zhang A, Lu F, Yang Y, Ren X, Li Z, Zhang W (2018) MicroRNA-217 overexpression induces drug resistance and invasion of breast cancer cells by targeting PTEN signaling. Cell Biol Int 42(10):1455. https://doi.org/10.1002/cbin.10506

Zhang D, Nichols HB, Troester M, Cai J, Bensen JT, Sandler DP (2020) Tea consumption and breast cancer risk in a cohort of women with family history of breast cancer. Int J Cancer 147(3): 876–886. https://doi.org/10.1002/ijc.32824. Epub 2020 Feb 4

Zhao T, Sun Q, Marques M, Witcher M (2015) Anticancer properties of Phyllanthus emblica (Indian gooseberry). Oxidative Med Cell Longev 2015:950890. https://doi.org/10.1155/2015/950890. Epub 2015 Jun 9